THE STRUCTURE OF LINEAR GROUPS

JOHN D. DIXON
Carleton University
Ottawa, Canada

VAN NOSTRAND REINHOLD COMPANY
LONDON
VNR
NEW YORK CINCINNATI TORONTO MELBOURNE

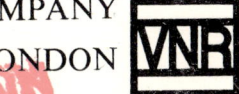

VAN NOSTRAND REINHOLD COMPANY
Windsor House, 46 Victoria Street, London S.W.1

INTERNATIONAL OFFICES
New York Cincinnati Toronto Melbourne

Library of Congress Catalog Card No. 73-160197
ISBN 0 442 02149 6

First published 1971

Printed in Great Britain by
Butler & Tanner Ltd
Frome and London

Contents

Acknowledgements

These notes were prepared over a period of two years, begun at the University of New South Wales and finished at Carleton University. I am indebted to both mathematics departments for their support in many ways. As well as my colleagues in these departments I owe a special acknowledgement to L. G. Kovacs at the Australian National University; although he has not read the manuscript, he helped me at a number of crucial points with generous advice and gentle criticism. Finally, I wish to thank Gill Chater, Jan Senior and Lynn McClelland who cheerfully and competently converted my handwriting into beautiful type.

1971 J. D. Dixon

Introduction

§0.1 The object of these notes is to study the structure of linear groups (or equivalently, matrix groups). The kind of results we are looking for are theorems which relate the linear structure (the degree of the group and the underlying field or ring) to the group theoretic structure (existence of normal subgroups, order of subgroups, solvability, etc.). This is similar to what we hope to get from a study of linear representations of groups, but the situation is different because: (1) we only start with a single faithful representation, and (2) we often consider infinite groups of a type not considered in representation theory.

The theory of linear groups began in 1876 with a paper by F. Klein on invariants. It was significantly extended by C. Jordan, who used it to study similar problems. The theory developed rapidly in the quarter-century around 1900 as further applications, such as the classifications of the crystallographic groups, arose; the names of Frobenius, Schur, Blichfeldt and Burnside will regularly occur throughout our work. This might be described as the classical period. Although the theory was used for studying finite groups, the linear groups considered were almost always over subfields of the complex numbers; some work by L. E. Dickson was an exception.

The modern period began around 1950 when it was discovered how linear groups arise in the analysis of the structure of abstract groups. An example is as follows. Let N be a normal subgroup in a group G. Then the centralizer $C_G(N)$ is normal in G and $G/C_G(N)$ is isomorphic to a subgroup of the group Aut N of automorphisms of N. If N is an elementary abelian p-group of order p^n, it behaves like a vector space of dimension n over a field with p elements, and Aut N is isomorphic to the group of all invertible linear transformations on that space. This makes $G/C_G(N)$ isomorphic to a group of linear transformations; what we call a linear group. This idea seems to have been first exploited in the infinite case, but most applications today are to finite groups. In these applications the fields are usually of characteristic $\neq 0$. As well as these applications, interest has revived in some of the classical problems and their generalizations to more general fields.

These notes try to give an introduction to the subject together with references to the rapidly expanding literature. The prerequisites, outlined in the next section, might be found in good undergraduate courses. Surprisingly little theory of characters and linear representations is ever used, although it would make a helpful background. On the other hand, some of our results would give a novel approach to parts of representation theory.

Note. Each chapter concludes with a section of Notes and References designed to relate the contents of the chapter to its history and literature. Many sections include exercises which give examples, counter-examples, extensions and applications of the theory.

The relation of dependence between chapters is roughly as follows:

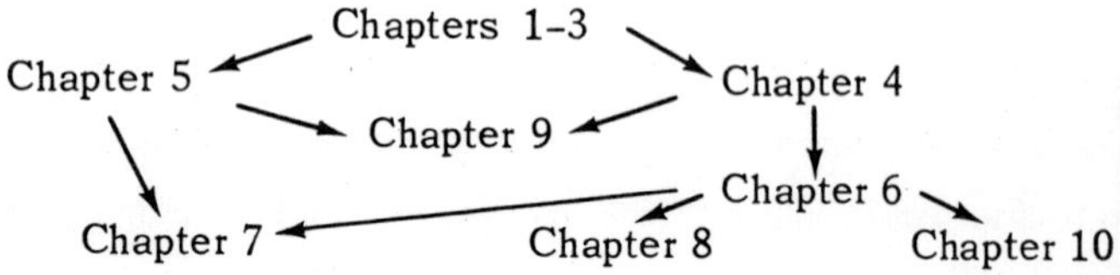

§0.2 Our prerequisites are as follows:

(1) *Linear Algebra.* Basic theorems including the theorem on the Jordan canonical form.

(2) *Group Theory.* Basic theorems including permutation groups, Sylow theorems and nilpotent and solvable groups. We also use the concept of an operator group and the notions of normal series, characteristic series and the Jordan-Hölder theorem for such a group. In Chapter 5 we need elementary character theory including the orthogonality relations between the irreducible characters of a finite group.

(3) *Commutative Algebra.* We shall deal with polynomial rings and ideals and use the Hilbert basis theorem. In field theory we need the ideas of field extension, algebraic closure, transcendence basis, finite normal extensions and Galois theory (for the separable case). We note two theorems especially:

If F is a field, and G is a finite subgroup of the multiplicative group of non-zero elements of F, then G is cyclic.

If F is a field with algebraic closure $\overline{F}$, then each finite extension E of F is isomorphic to subfield of $\overline{F}$ of the same degree over F.

In parts of Chapters 3 and 5 we also use a little algebraic number theory.

§0.3 The following notation will be used:

$\|H\|$, $\|G:H\|$	order, index of a subgroup H in a group G
$N_G(S)$, $C_G(S)$	normalizer, centralizer of S in G
$Z(G)$	center of G
$\langle S \rangle$	subgroup generated by S

$[x, y]$	$x^{-1}y^{-1}xy$
$[A, B]$	group generated by $[a,b]$ $(a \in A,\ b \in B)$
$G^{(i)}$	ith derived group
$\gamma_i(G), Z_i(G)$	terms in lower, upper central series
$H \lhd G$	H is a normal subgroup of G
Fit G, $\Phi(G)$	Fitting subgroup, Frattini subgroup
$\oplus$, x	direct sum, direct product
tr, det	trace, determinant
Z, **Q**, **C**	ring of integers, rational field, complex field
$[E : F]$	degree (of a field extension)
Gal(E/F)	Galois group

§0.4 *Notes and References.*

The material referred to in §0.2 is adequately covered by the following:

Curtis, C. W. and Reiner, I. *Representation Theory of Finite Groups and Associative Algebras,* Interscience, New York (1962).

Hall, M. *Theory of Groups,* Macmillan, New York (1959).

Huppert, B. *Endliche Gruppen,* vol. 1, Springer, Berlin (1967).

Kurosh, A.G. *Theory of Groups,* Chelsea, New York (1956).

Scott, W. R. *Group Theory,* Prentice-Hall, New Jersey (1964).

van der Waerden, B. L. *Modern Algebra,* Ungar, New York (1949).

Lang, S. *Algebra,* Addison-Wesley, Massachusetts (1967).

Classical references which include material on linear groups are:

Burnside, W. *Theory of Groups of Finite Order* (2nd ed. reprint), Dover, New York (1955).

Speiser, A. *Die Theorie der Gruppen von endlicher Ordnung,* Berlin (1937).

Also Chapter 10 of my own book gives an elementary introduction to linear groups.

Dixon, J. D. *Problems in Group Theory,* Blaisdell, Massachusetts (1967).

We might also mention

Wolf, J. A. *Spaces of Constant Curvature,* McGraw-Hill, New York (1967).

This book gives a detailed analysis of finite linear groups with fixed point free elements, in a rather unexpected setting.

CHAPTER 1

Some of the Classical Linear Groups

§1.1 Some classes of linear groups are very naturally defined in terms of their algebraic or geometric invariants. These are the so-called classical groups whose structure has been carefully studied and is now pretty well known. Since several books deal extensively with these groups (see §1.7), we shall only give a brief introduction to some of them here.

§1.2 deals with the general and special linear groups. §1.3 looks at some special classes of matrix groups and this leads into §1.4 where we look at the Sylow subgroups of the general linear groups. §1.5 deals with the symplectic groups which we shall need later on, and we conclude in §1.6 with some comments on the other classical groups. For notes and references see §1.7.

§1.2 Let V be a vector space of dimension n over a field F. (*Note:* all vector spaces and modules considered in these lectures are finite dimensional.) We write $\mathrm{Hom}_F(V, V)$ or sometimes just $\mathrm{Hom}(V, V)$ to denote the ring of all linear operators of V into itself. From elementary linear algebra we know that for any basis $v_1, \ldots, v_n$ for V over F there is an isomorphism $t \mapsto [\tau_{ij}]$ of $\mathrm{Hom}(V, V)$ onto the ring $M(n, F)$ of all $n \times n$ matrices over F where the $n \times n$ matrix $[\tau_{ij}]$ is defined by

$$v_i t = \sum_{j=1}^{n} \tau_{ij} v_j \quad (i = 1, \ldots, n).$$

Furthermore, if a different basis $u_i, \ldots, u_n$ is chosen for V, then the matrix $[\tau'_{ij}]$ corresponding to t over this basis satisfies

$$[\tau'_{ij}] = [\gamma_{ij}][\tau_{ij}][\gamma_{ij}]^{-1} \tag{1.2.1}$$

where $u_i = \Sigma\, \gamma_{ij} v_j$ $(i = 1, \ldots, n)$. The *units* of Hom(V, V), that is the invertible linear transformations on V, form a group under the ring multiplication called the *general linear group* $GL(V)$ (or sometimes written $GL_F(V)$). For any basis of V the set of matrices corresponding to $GL(V)$ forms the group of units of $M(n, F)$ called the *general linear group* $GL(n, F)$.

Because (1.2.1) holds, we can define unambiguously the determinant, characteristic polynomial, and trace of a linear operator t on V as the corresponding functions on the matrix $[\tau_{ij}]$ corresponding to t over any basis for V. In particular, the *special linear groups* $SL(V)$ and $SL(n, F)$ are the subgroups of $GL(V)$ and $GL(n, F)$, respectively, consisting of all elements with determinant 1.

A subgroup of $GL(V)$ is called a *linear group of degree n over the field F*. A subgroup of $GL(n, F)$ is called a *matrix group of degree n over F*. Because $GL(V) \simeq GL(n, F)$ any result about a linear group can be interpreted as a result about a matrix group and conversely. The object of these lectures is to discuss structural properties of linear groups (equivalently, matrix groups), particularly those group theoretic properties which can be deduced from a knowledge of the degree n and perhaps some knowledge about F. Sometimes it is easier to formulate the results or proofs in terms of linear groups, and sometimes in terms of matrix groups. We shall use the more convenient formulation, but it should be noted that each result has its alternative form (and that may be how it will be used later on). It is a useful exercise to give the alternative formulation of each of the theorems.

Let F be a finite field of order q; so q is a prime power and F is characterized up to isomorphism by q. In this case we write $GL(n, q)$ and $SL(n, q)$ in place of $GL(n, F)$ and $SL(n, F)$.

THEOREM 1.2. $|GL(n, q)| = q^{n(n-1)/2} \prod_{i=1}^{n} (q^i - 1) = (q-1)|SL(n, q)|$.

Proof. We have to count the number of ways of choosing a non-singular $n \times n$ matrix, or equivalently a matrix whose rows are linearly independent. The first row may be any non-zero row and so can be chosen $q^n - 1$ ways. After the first i rows have been chosen (linearly independently), the $(i+1)$st row must be chosen to be linearly independent of the first i rows, so it may be chosen $q^n - q^i$ ways. Hence, altogether the number of ways of choosing a non-singular matrix is

$$(q^n - 1)(q^n - q) \dots (q^n - q^{n-1})$$
$$= q^{n(n-1)/2} \prod_{i=1}^{n} (q^i - 1).$$

This proves the assertion for $|GL(n, q)|$. Since F has $q - 1$ non-zero elements, the homomorphism of $GL(n, q)$ defined by $x \mapsto \det x$ has image of order $q - 1$ and kernel $SL(n, q)$; hence $|GL(n, q)| = (q-1)|SL(n, q)|$. ■

Exercises. 1. Let E be a field extension of the field F. Then there is a natural isomorphism of $GL(n, F)$ onto a subgroup of $GL(n, E)$. In this way we may consider any matrix group of degree n over F as a matrix group of degree n over E.

2. The construction corresponding to Exercise 1 for linear groups is as follows. If V is a vector space of dimension n over F, then the tensor product $V^E = V \otimes_F E$ is a vector space of dimension n over E. There is a natural embedding of $GL(V)$ into $GL(V^E)$.

3. Show that the center of $GL(V)$ is the group of *scalars:* $\{a1 | a \in F,\ a \neq 0\}$ (here 1 is the identity of $GL(V)$).

4. If G is a finite subgroup of $GL(V)$, show that $e = \sum_{x \in G} x \in \mathrm{Hom}(V, V)$ satisfies $e^2 = |G| e$.

5. Let V be a vector space over an algebraically closed field F. Let S be the set of scalars in $GL(V)$. For any subgroup G of $GL(V)$, show that $G_0 = GS \cap SL(V)$ has the properties: $GS = G_0 S$ and $G/G \cap S \simeq G_0/G_0 \cap S$.

§1.3 There are some important matrix groups which have no natural analogue as linear groups. We discuss some of them in this section.

We first extend the definition of $M(n, R)$ to include the case of all $n \times n$ matrices over a ring R (all rings which we consider will have a unity element 1). This corresponds to the ring $\mathrm{Hom}_R(M, M)$ of all homomorphisms of a free module M of rank n over R into itself. The group of units of $M(n, R)$ is the general linear group $GL(n, R)$. In particular, in the case that R is commutative, it follows from Cramer's rule for inverting a matrix that: x_1 $M(n, R)$ is in $GL(n, R) \Longleftrightarrow \det x$ is a unit in R.

A matrix $x \in M(n, R)$ is (lower) *triangular* if it has the form

$$\begin{bmatrix} \xi_{11} & 0 & \cdots & 0 \\ & \xi_{22} & \cdots & 0 \\ & & \ddots & \vdots \\ & * & & 0 \\ & & & \xi_{nn} \end{bmatrix}$$

where the $*$ denotes unspecified entries below the main diagonal. Clearly the set of all triangular matrices is a subring of $M(n, R)$, and the set of triangular matrices in $GL(n, R)$ is a subgroup $TL(n, R)$ called the *triangular group*. The matrices in $TL(n, R)$ with diagonal entries all form the *special triangular group* $STL(n, R)$. In the case $x \in M(n, R)$ and all off-diagonal entries are zero, x is *diagonal*. The diagonal matrices in $GL(n, R)$ form the *diagonal group* $\mathrm{Diag}(n, R)$; diag $(\xi_{11}, \ldots, \xi_{nn})$ will denote a diagonal matrix whose diagonal entries are $\xi_{11}, \ldots, \xi_{nn}$. Sometimes we shall use the notation $x = \mathrm{diag}\,(x_1, \ldots, x_s)$ to denote a *block diagonal matrix*

$$\begin{bmatrix} x_1 & & & & \\ & x_2 & & 0 & \\ & & \cdot & & \\ & 0 & & \cdot & \\ & & & & x_s \end{bmatrix}$$

where each x_i is a square block of degree r_i, say, and all other entries are zero. A *monomial* matrix is a matrix in $GL(n, R)$ with exactly one non-zero entry (necessarily a unit in R) in each row and each column; if all non-zero entries are 1, then it is called a *permutation* matrix. It is readily verified that the set of all monomial matrices forms a subgroup $\mathrm{Mon}(n, R)$ of $GL(n, R)$, and the permutation matrices form a subgroup $\mathrm{Perm}(n, R)$. Moreover, $\mathrm{Mon}(n, R) = \mathrm{Diag}(n, R)\,\mathrm{Perm}(n, R)$, $\mathrm{Diag}(n, R) \cap \mathrm{Perm}(n, R) = 1$ and $\mathrm{Perm}(n, R) \simeq S_n$ the symmetric group.

Some of these concepts carry over in part to a vector space formulation. Let V be a vector of dimension n over a field F. Then $x \in GL(V)$ is *unipotent* if x has its n eigenvalues all equal to 1, or equivalently $(x-1)^n = 0$. If F is algebraically closed, then $x \in GL(V)$ is *semisimple* if, over some basis for V, x corresponds to a diagonal matrix (this happens exactly when the subalgebra $F(x) \subseteq \mathrm{Hom}(V, V)$ is a semisimple (associative) algebra). The Jordan canonical form theorem shows that each $x \in \mathrm{Hom}(V, V)$ is a product of mutually commuting x_1, x_2 where x_1 is semisimple and x_2 is unipotent. For further details see §8.6.

The proof of the following is left as an exercise.

LEMMA 1.3. *Let F be a field and F^* the multiplicative group of non-zero elements of F. Then:*

(1) $\mathrm{Diag}(n, F) \simeq F^* \times \ldots \times F^*$ *(n times).*

(2) *The derived group of $TL(n, F)$ is $STL(n, F)$.*

(3) *$STL(n, F)$ is nilpotent of class $n-1$ and solvable of length ℓ where $2^{\ell-1} < n \leqslant 2^{\ell}$.*

(4) *$TL(n, F)$ is the normalizer of $STL(n, F)$ in $GL(n, F)$.* ∎

Exercises. 1. Let V be a vector space of dimension n over a field F of characteristic $p > 0$. Show that $x \in GL(V)$ is unipotent $\Longleftrightarrow$ x is a p-element. Hence show that if x has order p^m then $p^{m-1} \leqslant n$. [*Hint:* Over a field of characteristic p, $(X^{p^k} - 1) = (X - 1)^{p^k}$.]

2. Show that $SL(n, F)$ is generated by the set of all elements $1 + ae_{ij}$ ($a \in F$; $i \neq j$; $i, j = 1, 2, \ldots, n$) where e_{ij} is the matrix with (i, j)th entry 1 and all other entries 0. Thus $SL(n, F)$ is generated by p-elements when char $F = p > 0$.

§1.4. *The Sylow subgroups of $GL(n, q)$.* Let F be a finite field of order q (where q is a prime power). Then, by Theorem 1.2, $GL(n, F) = GL(n, q)$ has order $q^{n(n-1)/2} \prod_{i=1}^{n} (q^i - 1)$. The present section deals with the structure of the Sylow p-groups of $GL(n, q)$ and the way that they are embedded in $GL(n, q)$. Naturally the cases where q is a power of p and where q is not a power of p differ sharply.

First suppose that q is a power of p. Then $q^{n(n-1)/2}$ is the largest power of p dividing $|GL(n, q)|$ and this is precisely the order of $STL(n, q)$. Thus $STL(n, q)$ is a Sylow p-group of $GL(n, q)$ and by the Sylow theorems we know that each p-subgroup of $GL(n, q)$ is conjugate to a subgroup of $STL(n, q)$. By Lemma 1.3, $TL(n, q)$ is the normalizer of $STL(n, q)$, and so $GL(n, q)$ has

$$|GL(n, q) : TL(n, q)| = \left\{ \prod_{i=1}^{n} (q^i - 1) \right\} / (q-1)^n$$
$$= \prod_{i=2}^{n} (1 + q + \ldots + q^{i-1})$$

Sylow p-groups. We summarize these results.

THEOREM 1.4A. *Let q be a power of p. Then $P = STL(n, q)$ is a Sylow p-group of $GL(n, q)$, and $TL(n, q) = N(P)$, the normalizer* of P in GL n, q . Each p-subgroup of *G is conjugate in G to a subgroup of P, and so has nilpotency class $\leqslant n - 1$ (Lemma 1.3). G has $\prod_{i=2}^{n} (1 + q + \ldots + q^{i-1})$ Sylow p-groups.* ■

Now consider the case where q is not a power of p and where $p \neq 2$. We first calculate the order of the Sylow p-groups in $GL(n, q)$. Define e as the smallest integer > 0 such that $q^e \equiv 1 \pmod p$, and write $q^e = 1 + mp$ $(r \geqslant 1,\ p \nmid m)$. If $t = t'p^s$ $(s \geqslant 0,\ p \nmid t')$, then since $p \neq 2$ the binomial theorem shows that the largest power of p dividing $q^{te} - 1 = (1 + mp^r)^t - 1$ is p^{r+s}. Define $d = [n/e]$ (the largest integer $\leqslant n/e$) and write $d = d_0 + d_1 p + \ldots + d_k p^k$ where $0 \leqslant d_i < p$. The factors in the order of $GL(n, q)$ (see Theorem 1.2) which are divisible by p are $(q^e - 1), (q^{2e} - 1), \ldots, (q^{de} - 1)$; from what we have just shown exactly $[d/p^s]$ of these are divisible by p^{r+s} $(s = 0, 1, \ldots)$. Hence the largest power of p dividing $|GL(n, q)|$ is p^h where

$$h = dr + [d/p] + [d/p^2] + \ldots$$
$$= dr + \sum_{i=1}^{k} d_i\,(1 + p + \ldots + p^{i-1}).$$

This gives the order of the Sylow p-groups of $GL(n, q)$.

Now suppose that for each $i \geqslant 0$ we are given a Sylow p-group P_i of $GL(ep^i, q)$; from above $|P_i| = p^{h_i}$ where

$$h_i = rp^i + (1 + p + \ldots + p^{i-1}) \quad (i = 0, 1, \ldots). \qquad (1.4.1)$$

We then define a p-group P in $GL(n, q)$ to consist of all matrices of the block diagonal form:

$$\text{diag}(1, \ldots, 1, x_{01}, \ldots, x_{0d_0}, \ldots, x_{k1}, \ldots, x_{kd_k}) \qquad (1.4.2)$$

where there are $(n - de)$ blocks 1 of degree 1, d_0 blocks $x_{0j} \in P_0, \ldots, d_k$ blocks $x_{kj} \in P_k$. Clearly P is a direct product of d_0 copies of $P_0, \ldots, d_k$ copies of P_k. Since $h = \sum_{i=1}^{k} d_i h_i$ from (1.4.1) and the definition of h, P has order p^h and so is a Sylow p-group of $GL(n, q)$. Summarizing we have the following result.

THEOREM 1.4B. *Let q be a prime power and let p be an odd prime not dividing q. Define e as the smallest integer > 0 such that $q^e \equiv 1 \pmod p$ and let p^r be the largest power of p dividing $q^e - 1$. For any integer $n \geqslant 1$, define $d = [n/e]$ and put $d = d \;\; + d\; p + \ldots +$ $d_k p^k$ $(0 \leqslant d_i < p)$. Then the set of all matrices of the form* (1.4.1) *forms a Sylow p-group of $GL(n, q)$. Thus P is isomorphic to a direct product of d_0 copies of P_0, d_1 copies of $P_1, \ldots,$ and d_k copies of P_k where P_i is a Sylow p-group of $GL(ep^i, q)$. Moreover, P_0 is a cyclic group of order p^r, and for $i \geqslant 1$, P_i may be taken as the group generated by the matrices (of degree ep^i):*

$$\begin{bmatrix} 0 & 1 & \cdots & 0 \\ 0 & 0 & & 0 \\ \vdots & & & \\ 0 & 0 & & 1 \\ 1 & 0 & & 0 \end{bmatrix} \quad \textit{and} \quad \operatorname{diag}(x, 1, \ldots 1), \tag{1.4.3}$$

where the 1'*s and* 0'*s are of degree* ep^{i-1} *and* x *runs over* P_{i-1}.

Proof. The first part of the theorem is proved above so it only remains to prove the assertions about P_0 and P_i $(i \geqslant 1)$. Let E be a field extension of degree e over the field F of order q. Since E is a vector space of dimension e over F, the group $GL_F(E) \simeq GL(e, q)$. Now $GL_F(E)$ contains the group of all non-zero right multiplications $\bar{\xi} : \eta \mapsto \eta\xi$ $(\eta \in E)$ where $\xi \in E$, $\xi \neq 0$. This latter group is isomorphic to the multiplicative group E^* of non-zero elements of E. But E^* is cyclic of order $q^e - 1$ (see §0.2) and p^r (the order of a Sylow p-group of $GL(n, q)$) divides $q^e - 1$. Thus a Sylow p-group of $GL(n, q)$ is isomorphic to a Sylow p-group of E^*, and hence P_0 is a cyclic group. Finally, to show that the matrices (1.4.3) generate a Sylow p-group of $GL(ep^i, q)$ we merely have to compare orders. Indeed the block diagonal matrices (1.4.3) generate a subgroup of index p and order $|P_{i-1}|^p$. Thus, all together the matrices (1.4.3) generate a group of order $p\,|P_{i-1}|^p = pp^{h_{i-1}p} = p^{h_i} = |P_i|$ by (1.4.1). ∎

Exercises. 1. If q is a power of p show that the largest order of any abelian p-subgroup of $GL(n, q)$ is q^f where $f = [n^2/4]$. There always exists an abelian group of this order, namely the group of all matrices of the form

$$\begin{bmatrix} 1 & 0 \\ x & 1 \end{bmatrix}$$

where x is an $n/2 \times n/2$ block if n is even, or an $(n-1)/2 \times (n+1)/2$ block if n is odd.

2. Find the number of Sylow p-groups of $GL(2, q)$ in the case where q is not a power of p.

§1.5 *The symplectic groups.* These groups will play an important part when we examine the structure of solvable linear groups. They are typical examples of the classical groups, and are constructed by first introducing a kind of 'distance' function into the underlying vector space, and then taking all 'distance-preserving' automorphism of the space.

DEFINITION. Let V be a vector space of dimension n over a field F. An *alternating form* on V is a function $f : V \times V \to F$ such that for all $u, v, w \in V$ and $\alpha, \beta \in F$:

(i) $f(\alpha u + \beta v, w) = \alpha f(u, w) + \beta f(v, w)$ and
$f(w, \alpha u + \beta v) = \alpha f(w, u) + \beta f(w, v)$ (bilinearity).

(ii) $f(u, u) = 0$.

We say f is *non-degenerate* if

(iii) for each $u \neq 0$, there exists v such that $f(u, v) \neq 0$.

We shall call the pair (V, f) a *sympletic space*, and say it is *non-degenerate* if f is non-degenerate.

Note. From (i) and (ii) we get $0 = f(u + v, u + v) - f(u, u) - f(v, v) = f(u, v) + f(v, u)$. Therefore

(ii)′ $$f(u, v) = -f(v, u).$$

Conversely, if char $F \neq 2$, then (i) and (ii)′ imply (ii).

Let (V, f) be a symplectic space. For any subset S of V we define the *perpendicular space* $S^{\perp} = \{v \in V \mid f(u, v) = 0 \text{ for all } u \in S\}$.

It follows from (i) that $S^{\perp}$ is a subspace of V; and $V^{\perp} = 0 \Longleftrightarrow f$ is non-degenerate.

We call two vectors $u, v \in V$ a *hyperbolic pair* if $f(u, v) = 1$; the two-dimensional subspace they span is then called a *hyperbolic plane.*

LEMMA 1.5. *Let (V, f) be a non-degenerate symplectic space of dimension n over a field F. Then:*

(1) *For each subspace U of V, if $U \cap U^{\perp} = 0$ then $V = U \oplus U^{\perp}$ (direct sum).*

(2) *$V = H_1 \oplus \ldots \oplus H_m$ where each H_i is a hyperbolic plane, and $H_i \subseteq H_j^{\perp}$ for all $i \neq j$. In particular, $n = 2m$ is even.*

Proof. (1) If $u_1, \ldots, u_k$ is a basis for U, then $U^{\perp}$ is the set of all vectors $v \in V$ satisfying the k linear conditions $f(u_i, v) = 0$ $(i = 1, \ldots, k)$. Hence, by elementary facts about linear equations, $\dim U^{\perp} \geqslant \dim V - k = \dim V - \dim U$. Hence, if $U \cap U^{\perp} = 0$, a comparison of dimensions shows that $V = U \oplus U^{\perp}$.

(2) If $V = 0$ the result is trivial so let $n \geqslant 1$. Choose $u_1 \neq 0$ in V. Since f is non-degenerate, we can find $v_1 \in V$ such that $f(u_1, v_1) \neq 0$. Multiplying v_1 by a suitable scalar and using (i) above we may suppose $f(u_1, v_1) = 1$. Let H_1 be the hyperbolic plane with u_1, v_1 as basis. Then for all $\alpha u_1 + \beta v_1 \in H_1$ $(\alpha, \beta \in F)$ we have $f(u_1, \alpha u_1 + \beta v_1) = \beta$ and $f(v_1, \alpha u_1 + \beta v_1) = -\alpha$; thus $H_1 \cap H_1^{\perp} = 0$. Hence from (1) we conclude $V = H_1 \oplus H_1^{\perp}$. The restriction $\bar{f}$ of f to $H_1^{\perp} \times H_1^{\perp}$ is an alternating form on $H_1^{\perp}$. Moreover $\bar{f}$ is non-degenerate. Indeed, if $w \neq 0$ is in $H^{\perp}$, then we can find $u \in H$, $v \in H^{\perp}$ such that $f(w, u+v) \neq 0$ because f is non-degenerate; hence $0 \neq f(w, u) + f(w, v) = \bar{f}(w, v) = f(w, v)$ so $\bar{f}$ is non-degenerate. Thus $(H^{\perp}, \bar{f})$ is a non-degenerate symplectic space of dimension $n-2$. The result now follows by an induction on the dimension. ■

Let (V, f) be a non-degenerate symplectic space of dimension $2m$. A *hyperbolic basis* for V is a basis $u_1, v_1, \ldots, u_m, v_m$ such that u_i, v_i is a hyperbolic pair $(i = 1, \ldots, m)$ and $f(u_i, u_j) = f(u_i, v_j) = f(v_i, v_j) = 0$ for all $i \neq j$. By Lemma 1.5 (2) each non-degenerate symplectic space has a hyperbolic basis (take u_i, v_i as a hyperbolic pair spanning H_i).

Now suppose that (V, f) and (W, g) are two symplectic spaces over the same field F. An *isometry* $t : (V, f) \to (W, g)$ is an invertible linear transformation of V onto W such that $f(u, v) = g(ut, vt)$ for all $u, v \in V$. If an isometry exists, we call (V, f) and (W, g) *isometric;* such spaces have the same dimension because an isometry is invertible. If (V, f) and (W, g) are both non-degenerate, then an isometry will map a hyperbolic basis of the first space onto a hyperbolic basis of the second space. The converse is also true: if $u_1, v_1, \ldots, u_m, v_m$ and $u_1', v_1', \ldots, u_m', v_m'$ are hyperbolic spaces for two non-degenerate spaces (V, f) and (W, g), then the mapping $t : V \to W$ defined by

$$\sum_{i=1}^{m} (\alpha_i u_i + \beta_i v_i)\, t \overset{\text{def}}{=} \sum_{i=1}^{m} (\alpha_i u_i' + \beta_i v_i')$$

(for all $\alpha_i, \beta_i \in F$) is the unique isometry taking the first basis onto the second basis. In particular, any two non-degenerate symplectic spaces of the same dimension over the same field are isometric.

DEFINITION. Let (V, f) be a non-degenerate symplectic space of dimension $2m$ over a field F. The *symplectic group* $Sp_f(V)$ is the subgroup of $GL(V)$ consisting of all isometries of (V, f) into itself.

Note. If (W, g) is also a non-degenerate symplectic space of dimension $2m$ over F, then as we have seen, there is an isometry $t : (V, f) \to (W, g)$. We then have $Sp_g(W) = t^{-1}Sp_f(V)t$. In particular, the symplectic groups are characterized up to isomorphism by the dimension $2m$ and the field F, and any two symplectic groups over the same space V are conjugate in $GL(V)$. We shall use the notations $Sp(V)$ or $Sp(2m, F)$ when we do not need to specify the particular symplectic space involved. When F is finite of order q we shall also write $Sp(2m, q)$ for $Sp(2m, F)$.

THEOREM 1.5. $|Sp(2m, q)| = q^{m^2} \prod_{i=1}^{m} (q^{2i} - 1)$.

Proof. Let (V, f) be a non-degenerate symplectic space of dimension $2m$ over a field of order q, and let $u_1, v_1, \ldots, u_m, v_m$ be a fixed hyperbolic basis. As we have seen above, to each element $t \in Sp_f(V)$, there is a hyperbolic basis $u_1t, v_1t, \ldots, u_mt, v_mt$ of (V, f) and conversely each hyperbolic basis of (V, f) arises in this way. Thus $|Sp(2m, q)| = |Sp_f(V)|$ equals the number of hyperbolic bases for (V, f). We shall use induction on m to count the number of ways of choosing hyperbolic bases for (V, f) (compare with the proof of Lemma 1.5).

The first vector u_1 of a hyperbolic basis may be chosen in $q^{2m} - 1$ ways as any non-zero vector. Then $\{u_1\}^{\perp}$ is the set of vectors in V satisfying one non-trivial linear condition and so $\dim \{u_1\}^{\perp} = \dim V - 1 = 2m - 1$. Thus $q^{2m} - q^{2m-1}$ vectors $v \in V$ satisfy $f(u_1, v) \neq 0$. Since there are $q - 1$ non-zero scalars, this means that there are $(q^{2m} - q^{2m-1})/(q - 1) = q^{2m-1}$ vectors v such that $f(u_1, v) = 1$. Hence there are $(q^{2m} - 1)q^{2m-1}$ ways of choosing a hyperbolic pair u_1, v_1 in V. Then the remaining vectors for our basis must be chosen as a hyperbolic basis for the non-degenerate symplectic space $(H_1^{\perp}, \bar{f})$ where $\bar{f}$ is the restriction of f to $H_1^{\perp} \times H_1^{\perp}$ (see the proof of Lemma 1.5). This can be done in $|Sp(2m-2, q)|$ ways. Hence by the induction hypothesis

$$|Sp(2m, q)| = (q^{2m} - 1)q^{2m-1}\,|Sp(2m-2, q)|$$

$$= q^{m^2} \prod_{i=1}^{m} (q^{2i} - 1). \blacksquare$$

Exercises. 1. If (V, f) is a non-degenerate symplectic space of dimension 2, show that $Sp_f(V) = SL(V)$.

2. Let $u_1, v_1, \ldots, u_m, v_m$ be a hyperbolic basis for a non-degenerate symplectic space (V, f) over a field F. Show that over this basis $Sp_f(V)$ corresponds to the group of all $x \in GL(2m, F)$ such that $x\,e\,x^T = e$ (x^T is the transpose of x) where $e = \operatorname{diag}(e_0, e_0, \ldots, e_0)$ with

$$e_0 = \begin{bmatrix} 0 & 1 \\ -1 & 0 \end{bmatrix}.$$

3. Let (V, f) be a symplectic space with f degenerate and put $W = V^{\perp}$. Show that we can define a non-degenerate alternating form $\bar{f}$ on V/W by $\bar{f}(u + W, v + W) = f(u, v)$ (check $\bar{f}$ is well-defined). Prove that the group of all isometries of (V, f) into itself is isomorphic to $Sp_{\bar{f}}(V/W) \times GL(W)$.

4. Let E be a field extension of degree n over a field F. Show that there is an isomorphism of $Sp(2m, E)$ onto a subgroup of $Sp(2mn, F)$. [*Hint:* A vector space of dimension $2m$ over E may also be considered as a vector space of dimension $2mn$ over F.]

§1.6 The groups $GL(V)$, $SL(V)$ and $Sp(V)$ are three of the classical groups. In a manner similar to the way in which $Sp(V)$ was defined, we can also define *unitary* and *orthogonal* groups which are analogous to the groups of unitary and orthogonal transformations of an inner product space over the complex numbers. One reason that the classical groups are important is because they are related to families of simple groups.

Although we shall not use the following result in these notes, it does play an important part in the structure theory of finite groups. Hence we state the theorem and refer to the literature quoted in §1.7 for a proof.

THEOREM 1.6. *The* projective special linear groups $PSL(n, F)$ *and the* projective symplectic groups $PSp(n, F)$ *are defined as the factor groups of* $SL(n, F)$ *and* $Sp(n, F)$ *over their respective centers (consisting of scalars). These projective groups are all simple for* $n \geqslant 2$ *except in the following cases:*

(1) $PSL(2, 2) = PSp(2, 2) = SL(2, 2) \simeq S_3$ *(the symmetric group of order 6).*

(2) $PSL(2, 3) = PSp(2, 3) \simeq A_4$ *(the alternating group of order 12).*

(3) $PSp(4, 2) = Sp(4, 2) \simeq S_6$ *(the symmetric group of order 720).* ■

§1.7 *Notes and References*

The results of this chapter are mostly classical, dating back at least as far as Dickson's book first published in 1901:

Dickson, L. E. *Linear Groups* (Reprint), Dover, New York, (1958).

A modern exposition of these results is given in

Dieudonné, J. *La géometrie des groupes classiques* (2nd ed.), Springer, Berlin (1963).

The discussion of symplectic groups in §1.5 is based on the approach in

Artin, E. *Geometric Algebra,* Interscience, New York (1957).

It may be contrasted in clarity with the matrix theoretic approach in Dickson's book. Incidentally, the term 'abelian linear group' is used in the latter book in place of 'symplectic group'; Weyl has also used the term 'complex group'. The proof of Theorem 1.6 may be found in Artin's book or Huppert's book (see §0.4). A further useful reference for the material in §1.5–1.6 is

Carter, R. W. Simple groups and simple Lie algebras, *J. Lond. Math. Soc.* **40**, 193–240 (1965)

The results of §1.4 are taken from the following papers:

Weir, A. J. Sylow *p*-subgroups of the general linear group over finite fields of characteristic *p*, *Proc. Am. Math. Soc.* **6**, 454–464 (1955)

—— Sylow *p*-subgroups of the classical groups over finite fields with characteristic prime to *p*, *Proc. Am. Math. Soc.* **6**, 529–533 (1955)

Further results are also given there. The case $p = 2$ is omitted in Theorem 1.4B; it may be found in:

Carter, R. and Fong, P. The Sylow 2-subgroups of the finite classical groups, *J. Algebra,* **1**, 139–151 (1964).

Exercise 1 of §1.4 is from an unpublished thesis of J. T. Goozeff. Information about Exercise 2 may be found in Huppert's book or Burnside's book (see §0.4).

CHAPTER 2

Reducibility and Complete Reducibility

§2.1. The concepts of reducibility and complete reducibility which we shall define below are fundamental in any discussion of linear groups. This chapter deals with basic theorems in this area, and includes results dating back to 1900 and before. First we have two criteria insuring complete reducibility, Clifford's theorem (§2.2) and Maschke's theorem (§2.3), and a method of obtaining completely reducible representations (§2.4). Then we examine the irreducible case and prove Schur's lemma (§2.6) and the Burnside-Frobenius-Schur theorem (§2.7) and give some applications. Finally, we consider what happens to irreducibility when the underlying field is extended (§2.10).

We now give the basic definitions. Let V be a vector space of dimension n over a field F, and let G be a subgroup of $GL(V)$. A subspace W of V is a *G-space* (or *invariant* under G) if $Wx \subseteq W$ for all $x \in G$. We say G is *irreducible* if the only G-spaces are 0 and V; otherwise G is *reducible*. We say G is *completely reducible* if there are minimal (non-zero) G-spaces $W_1, \ldots, W_s$ $(1 \leqslant s \leqslant n)$ such that $V = \bigoplus_{i=1}^{s} W$ (direct sum); in particular, an irreducible group is completely reducible.

Suppose W is a G-space of V. Then each $x \in G$ induces linear transformations $x|W$ and $x|V/W$ on W and the factor space V/W, respectively. Specifically: $w(x|W) = wx$ for all $w \in W$, and

$(v + W)(x|V/W) = vx + W$ for all $v + W \in V/W$. Clearly $x \mapsto x|W$ and $x \mapsto x|V/W$ are homomorphisms of G into $GL(W)$ and $GL(V/W)$, respectively; we write the images of these homomorphisms as $G|W$ and $G|V/W$.

Let W_1, W_2 be G-spaces of V. Then a linear mapping $t : W_1 \to W_2$ is a *G-homomorphism* if the following diagram commutes for all $x \in G$:

$$\begin{array}{ccc} W_1 & \xrightarrow{\ t\ } & W_2 \\ \downarrow x & & \downarrow x \\ W_1 & \xrightarrow{\ t\ } & W_2 \end{array}$$

The alternative point of view of representation theory is sometimes helpful. Given an (abstract) group G and a vector space V, then a *representation* of G over V is a homomorphism $\rho : G \to GL(V)$; ρ is *faithful* if its kernel is 1. Now the image G^ρ is a linear group on V, and we say (inexactly) that W is a G-space when it is a G^ρ-space, and call ρ irreducible, reducible, or completely reducible if G^ρ is. If $\rho : G \to GL(V)$ and $\sigma : G \to GL(W)$ are two representations of G, where V and W are vector spaces over the same field, then ρ is *equivalent* to σ if there is an invertible linear mapping $t : V \to W$ such that

$$\begin{array}{ccc} V & \xrightarrow{\ t\ } & W \\ \downarrow x^\rho & & \downarrow x^\sigma \\ V & \xrightarrow{\ t\ } & W \end{array}$$

commutes for all $x \in G$. In particular, if ρ and σ are equivalent, then $\dim V = \dim W$. The mapping t will be called a *G-isomorphism*.

Exercises. 1. Interpret all these terms for matrix groups.

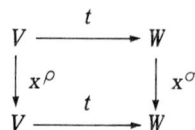

2. Show that a subgroup G of $GL(V)$ is completely reducible if and only if whenever W is a G-space, then there is a G-space U such that $V = U \oplus W$. Such a subspace U is called a *complementary* subspace of W.

3. Show that the relation of equivalence defined above for representations is really an equivalence relation (reflexive, symmetric and transitive).

§2.2. We begin with two criteria which imply complete reducibility. The first is due to A. H. Clifford (1937).

THEOREM 2.2. *Let G be an irreducible subgroup of $GL(V)$ where V is a vector space of dimension n, and let N be a normal subgroup of G. Then there are elements $y_1 = 1, \ldots, y_d$ in G such that*

$$V = Wy_1 \oplus \ldots \oplus Wy_d \tag{2.2.1}$$

where the Wy_i are minimal N-spaces of V. In particular, N is completely reducible, and the irreducible components $N|Wy_i$ $(i = 1, \ldots, d)$ are isomorphic linear groups of the same degree n/d.

Proof. Let W be a minimal N-space. For each $x \in N$, $y \in G$ we have $(Wy)x = (Wyxy^{-1})y = Wy$ because $yxy^{-1} \in N$. Hence Wy is an N-space and it is readily verified that it is a minimal N-space. Now choose $y_1 = 1, \ldots, y_d$ in G such that $Wy_1 + \ldots + Wy_d$ is a direct sum $V_1 = \bigoplus_i Wy_i$ and so that no larger set of elements of G have this property; obviously $d \leqslant n$. Then V_1 is an N-space, and for each $y \in G$, $V_1 \cap Wy = Wy$ or 0 because Wy is a minimal N-space. By the choice of d, this implies that $V_1 \cap Wy = Wy$; hence we have $Wy \subseteq V_1$ for all $y \in G$. But this implies that V_1 is a G-space and so $V_1 = V$

by the irreducibility of G. This proves (2.2.1). Finally, we note that mapping $x|Wy_i \leftrightarrow (y_j^{-1} y_i\, x y_i^{-1} y_j)|Wy_j$ is an isomorphism of $N|Wy_i$ onto $N|Wy_j$. ■

A subgroup H of G is called *subnormal* if there is a finite chain of subgroups

$$G = H_0 \supseteq H_1 \supseteq \ldots \supseteq H_k = H$$

with $H_i \lhd H_{i-1}$ $(i = 1, \ldots, k)$. An obvious induction on the length of such chains gives the following corollary.

COROLLARY 2.2. *Let G be a completely reducible linear group. Then every subnormal subgroup of G is completely reducible.* ■

§2.3. Our second result on complete reducibility is a generalization of a classical theorem of H. Maschke (1898).

THEOREM 2.3. *Let G be a subgroup of $GL(V)$ where V is a vector space over a field F, and let H be a subgroup of finite index h in G. Suppose:*

(1) char $F \nmid h$ *(in particular, this holds if* char $F = 0$*); and*

(2) *H is completely reducible.*

Then G is completely reducible.

Remark. The most important case ($H = 1$) is the original theorem of Maschke.

Proof. From Exercise 2 of §2.1 it is enough to show that if W is a G-space then it has a complementary G-space in V. Since H is completely reducible, there is an H-space U such that $V = U \oplus W$. Define $e \in \mathrm{Hom}(V, V)$ as the projection onto U along W : $(u + w)e = u$

for all $u \in U$, $w \in W$. Let $y_1, \ldots, y_h$ be a set of right coset representatives of H in G. Then define $f \in \mathrm{Hom}(V, V)$ by

$$f = \frac{1}{h} \sum_{i=1}^{h} y_i^{-1} e y_i.$$

This is possible because $1/h \in F$ by (1). Now note that, if x and y are in the same right coset of H in G, then $Vx^{-1}ex = Ux = Uy = Vy^{-1}ey$ because U is an H-space. Therefore Vf is independent of the coset representatives for H in G; we shall prove Vf is a G-space complementary to W in V.

First, for each $x \in G$, $y_1 x, \ldots, y_h x$ is a set of right coset representatives for H so

$$\begin{aligned} Vf &= V \cdot \frac{1}{h} \sum_{i=1}^{h} (y_i x)^{-1} e (y_i x) \\ &= Vx^{-1} \cdot \frac{1}{h} \sum_{i=1}^{h} y_i^{-1} e y_i \cdot x \\ &= Vx^{-1} f x = Vfx. \end{aligned}$$

Thus Vf is a G-space. Secondly, since $Wy = W$ for all $y \in G$, and $We = 0$, we conclude that $Wf = 0$ and so $w(1 - f) = w$ *for each* $w \in W$. However, the definition of e shows that $V(1 - e) = W$, so for all $v \in V$ and each i, $vy_i^{-1}(1 - e)y_i \in V(1 - e)y_i = W$; therefore $v(1 - f) \in W$. Hence we conclude that $W = V(1 - f)$ and so $V = Vf + V(1 - f) = Vf + W$. On the other hand, $vf \in W$ for some $v \in V$ implies $vf = vf(1 - f) \in V(1 - f)f = Wf = 0$; hence $Vf \cap W = 0$. Therefore $V = Vf \oplus W$ as required. ∎

Exercises. 1. Show that for any prime p, $STL(n, p)$ is not completely reducible when $n > 1$ (hence condition (1) of the theorem cannot be dropped).

2. Find a subgroup of $GL(2, p)$ which is completely reducible, but does not satisfy the conditions of Maschke's theorem.

3. Let G be a completely reducible linear group over a field F of characteristic 0. Show that each subgroup of finite index in G is completely reducible. What about the case where char $F > 0$? [*Hint:* Each subgroup of finite index contains a normal subgroup of finite index.]

2.4. If we are only interested in the group theoretic structure of a linear group, the following result is useful in reducing our considerations to completely reducible groups. Although it is almost trivial, this theorem proves to be a very useful tool.

THEOREM 2.4. *Let G be a subgroup of GL(V) where V is a vector space of dimension n over a field F. Let*

$$V = V_0 \supset V_1 \supset \dots \supset V_s = 0$$

be a G-composition series for V; that is, each V_i *is a G-space and each of the groups* $G|V_{i-1}/V_i$ $(i = 1, \dots, s)$ *is a non-trivial irreducible group. Then there is a completely reducible representation* $\rho : G \to GL(V)$ *such that each* V_i *is a* G^ρ*-space, and for each* $x \in G$ *we have* $x|V_{i-1}/V_i = x^\rho|V_{i-1}/V_i$ $(i = 1, \dots, s)$; *in particular*

$$\operatorname{tr} x^\rho = \sum_{i=1}^{s} \operatorname{tr}(x|V_{i-1}/V_i) = \operatorname{tr} x.$$

The kernel of ρ *is a group of unipotent elements in G, and indeed over a suitable basis for V,* ker ρ *corresponds to a subgroup of* $STL(n, F)$.

Proof. Choose any subspace W_i of V_{i-1} such that $V_{i-1} = W_i \oplus V_i$ $(i = 1, \dots, s)$. We then define ρ so that the obvious isomorphism $V_{i-1}/V_i \to W_i$ induces $x|V_{i-1}/V_i \leftrightarrow x^\rho|W_i$. Specifically, if $x \in G$ and $w \in W_i$ then we define $wx^\rho = w'$ where $w' \in W_i$ and $(w + V_i)x = w' + V_i$. Then $x^\rho|W_i \in GL(W_i)$ for $i = 1, \ldots, s$. Since $V = W_1 \oplus \ldots \oplus W_s$ we can extend the definition of x^ρ uniquely by linearity so that $x^\rho \in GL(V)$. It is now readily verified that $\rho : G \to GL(V)$ is representation of G with $\ker \rho = \{x \in G|\ x|V_{i-1}/V_i = 1$ for $i = 1, \ldots, s\}$. If we choose $w_{i1}, \ldots, w_{in,i}$ as a basis for W_i $(i = 1, \ldots, s)$ then, over the basis $w_{s1}, \ldots, w_{sn,s}, \ldots, w_{11}, \ldots, w_{1n,1}$ for V, $\ker \rho$ corresponds to a subgroup of $STL(n, F)$. ■

COROLLARY 2.4. *Let G be a subgroup of GL(V) where V is a vector space of dimension n over a field F. Then G has a normal subgroup N such that G/N is isomorphic to a completely reducible subgroup of GL(V) and N is nilpotent group of class* $\leqslant n - 1$. *If* char $F = p > 0$, *then N is a normal p-subgroup G.*

Proof. We take $N = \ker \rho$ in the Theorem. Then apply Lemma 1.3 and Exercise 1 of §1.3. ■

Note. See Theorem 2.8C for further results.

§2.5. The following results are due to A. I. Mal'cev (1941).

If G is a subgroup of $GL(V)$ where V is a vector space of dimension n over a field F, then FG will denote the subspace (in fact, subalgebra) of $\mathrm{Hom}_F(V, V)$ spanned by the elements of G; the dimension of FG over F is at most $n^2 = \dim \mathrm{Hom}_F(V, V)$.

THEOREM 2.5. *Let G be a subgroup of GL(V). If each finitely generated subgroup of G is completely reducible, then G is completely reducible.*

Proof. Choose $x_1, \ldots, x_m$ in G as a basis for FG over F. $H = \langle x_1, \ldots, x_m \rangle$ is a finitely generated subgroup of G, and so is completely reducible. Since $FH = FG$, each H-space is a G-space and so G is completely reducible. ∎

A group is called *locally finite* is each finitely generated subgroup is finite. Clearly a locally finite group is *periodic* (each element has finite order), but the converse is not generally true. However, we shall prove later (Theorem 9.2) that each periodic linear group is locally finite.

COROLLARY 2.5. *If G is a locally finite linear group over a field F, and each element of G has order prime to* char F, *then G is completely reducible.*

Remark. Conventially, the second condition is always satisfied when char $F = 0$.

Proof. From Maschke's theorem (§2.3) and the Theorem above. ∎

Note. The converse of Theorem 2.5 is false: a completely reducible linear group (for example, $GL(V)$) may have subgroups which are not completely reducible. However, there is a converse for solvable groups (see Theorem 6.6).

Exercise. Let G be a linear group in which each finitely generated subgroup is reducible. Prove that G is reducible.

§2.6. The following classical result of I. Schur (1905) is generally referred to as *Schur's lemma.*

THEOREM 2.6. *Let V and W be vector spaces over a field F and let $\rho : G \to GL(V)$ and $\sigma : G \to GL(W)$ be irreducible representations of a group G. Define L to be the set of all $t \in \mathrm{Hom}_F(V, W)$ such that*

$$x^{\rho}t = tx^{\sigma} \quad \text{for all } x \in G. \tag{2.6.1}$$

Then:

(1) $L = \{0\}$ *if* ρ *is not equivalent to* σ *(see* §2.1*).*

(2) *If* $\rho = \sigma$, *then* $L \subseteq \mathrm{Hom}(V, V)$ *is a division subring.*

(3) $L = F.1$ *(the set of all scalars in* $\mathrm{Hom}(V, V)$*) if* $\rho = \sigma$ *and* F *is algebraically closed.*

Proof. Let $t \in L$. We conclude from (2.6.1) that Vt is a G^{σ}-space of W; because σ is irreducible this means $Vt = 0$ or W. Similarly (2.6.1) shows that the kernel $K = \{v \in V \mid vt = 0\}$ of t is a G^{ρ}-space of V; because ρ is irreducible this means $K = 0$ or V. Thus either $t = 0$ ($Vt = 0$, $K = V$) or else t is an isomorphism of V onto W ($Vt = W$, $K = 0$). In the latter case t has an inverse and it is immediately verified that ρ is equivalent to σ. Since in case (2) L is clearly a subring of $\mathrm{Hom}(V, V)$, the assertions (1) and (2) follow. Finally, if F is algebraically closed and $\rho = \sigma$, then each $t \in L$ has at least one eigenvalue, say $\lambda \in F$. Since $t - \lambda 1 \in L$ and $t - \lambda 1$ is not invertible, (2) shows that $t - \lambda 1 = 0$. Thus $L = F \cdot 1$ and (3) is proved. ■

Interpreting (3) in the language of linear groups we get the following corollary.

COROLLARY 2.6. *Let* G *be an irreducible subgroup of* $GL(V)$ *where* V *is a vector space over an algebraically closed field* F*. Then the only elements of* $\mathrm{Hom}(V, V)$ *which commute with* G *are the scalars* $\lambda 1$ ($\lambda \in F$). *In particular, if* G *is abelian, then* G *is a group of scalars; therefore since* G *is irreducible it has degree* 1. ■

Exercises. 1. Let F be an arbitrary field, and let $G = TL(2, F)$. Show that the only elements of $M(2,F)$ which centralize G are scalars even though G is not completely reducible.

2. Let G be a completely reducible subgroup of $GL(V)$ where V is a vector space over an algebraically closed field F. If the only

elements of $\mathrm{Hom}(V, V)$ which centralize G are scalars, show that G is irreducible. (Partial converse of Corollary 2.6.)

3. Show that the condition that F be algebraically closed is necessary in part (3) of the Theorem.

4. Let G be an abelian completely reducible subgroup of $GL(V)$ where V is a vector space over an algebraically closed field. Show that, for a suitable basis of V, G corresponds to a group of diagonal matrices.

5. Let G be a finite p-subgroup of $GL(V)$ where V is a vector space of dimension $n > 1$ over an algebraically closed field. Show that G has a subgroup H of index p which is reducible. [*Hint:* We may suppose G is irreducible. Let $1 = Z_0 \subset Z_1 \subset \ldots$ be the upper central series for G. Choose $x \in Z_2$, $x \notin Z_1$ such that the element xZ_1 has order p in G/Z_1, and take $H = C_G(x)$.]

6. Let G be a finite subgroup of $\mathrm{Diag}(n, F)$ for some field F. Then G can be generated by $\leqslant n$ elements. [*Hint:* Finite subgroups of the multiplicative group F^* of F are cyclic (see §0.2).]

§2.7. The present section deals with key results due to I. Schur, G. Frobenius, and W. Burnside (about 1906) on properties of irreducible representations.

LEMMA 2.7. *Let $\sigma : G \to GL(V)$ be an irreducible representation of a group G on a vector space V over a field F. Let L be a non-zero subspace of the vector space* $\mathrm{Hom}(V, V)$ *over F such that L is minimal with respect to the condition $Lx^\sigma = L$ for all $x \in G$. Then there is an irreducible representation $\rho : G \to GL(L)$ defined by $x^\rho : u \mapsto ux^\sigma$ $(u \in L)$. Moreover there is some $v \in V$ such that the mapping $u \mapsto vu$ is a G-isomorphism from L onto V and so ρ and σ are equivalent representations; in particular,* $\dim L = \dim V$.

Proof. It is readily verified that ρ is a representation of G, and ρ is irreducible because L is minimal. Since σ is irreducible, the minimal G^σ-space vL of V equals 0 or V for each $v \in V$. Since $L \neq 0$, we can choose $v \in V$ so that $vL = V$. The mapping $\theta : u \mapsto vu \ (u \in L)$ is a G-homomorphism of L onto V. Since the kernel of θ is a G^ρ-space of L, and ρ is irreducible, $\ker \theta = 0$ or L. The latter is impossible so we conclude $\ker \theta = 0$ and θ is a G-isomorphism of L onto V. ∎

THEOREM 2.7A. *Let G be a group, F an algebraically closed field, and V_i a vector space of dimension n_i over F $(i = 1, \ldots, s)$. Suppose $\sigma_i : G \to GL(V_i)$ are pairwise inequivalent irreducible representations of G $(i = 1, \ldots, s)$. Then the only elements $c_i \in \mathrm{Hom}(V_i, V_i)$ such that*

$$\sum_{i=1}^{s} \mathrm{tr}(c_i x^{\sigma_i}) = 0 \quad \text{for all } x \in G$$

are all $c_i = 0$.

Proof. We proceed by induction on s.

CASE 1 $(s = 1)$. Write $\sigma_1 = \sigma$, $V_1 = V$ and $n_1 = n$. Suppose the assertion is false. Then

$$M = \{c \in \mathrm{Hom}(V, V) \mid \mathrm{tr}(cx^\sigma) = 0 \text{ for all } x \in G\}$$

is a non-zero subspace of $\mathrm{Hom}(V, V)$ with $My^\sigma = M$ for all $y \in G$. Let L be a non-zero subspace of M which is minimal with respect to the condition $Ly^\sigma = L$ for all $y \in G$. By Lemma 2.7, $\dim L = n$ and for some $v \in V$, $vL = V$. Let $u_1, \ldots, u_n$ be a basis for L over F; then $v_1 = vu_1, \ldots, v_n = vu_n$ is a basis for V. Now for each $w \in V$ we have $[a_{ij}] \in M(n, F)$ such that

$$wu_i = \sum_{j=1}^{n} a_{ij} v_j \quad (i = 1, \ldots, n).$$

Since F is algebraically closed, the matrix $[a_{ij}]$ has an eigenvalue $\lambda_w \in F$, say; and then the n vectors

$$(w - \lambda_w v)u_i = \sum_{j=1}^{n} (a_{ij} - \delta_{ij} \lambda_w)v_j \quad (i = 1, \ldots, n)$$

are linearly dependent in V. Since these vectors span a G^{σ}-space $(w - \lambda_w v)L$ of L we conclude from the irreducibility of σ that $(w - \lambda_w v)L = 0$. Changing the notation and writing λ_j for λ_{v_j} we have $(v_j - \lambda_j v)L = 0$ $(j = 1, \ldots, n)$. Therefore

$$v_j u_i = \lambda_j v_i \quad (i, j = 1, \ldots, n). \tag{2.7.1}$$

In particular, tr $u_i = \lambda_i$ (consider the matrix of u_i over the basis $v_1, \ldots, v_n$). Because $L \subseteq M$, $0 = \mathrm{tr}(u_i \cdot 1) = \lambda_i$ for each i. Hence (2.7.1) shows that each $u_i = 0$. This contradicts the choice of $u_1, \ldots, u_n$ as a basis for L. Hence we conclude $M = 0$ and the assertion is proved in this case.

CASE 2 $(s > 1)$. Suppose the assertion is false. Then, as in Case 1, we can find a subspace L of the vector space $\mathrm{Hom}(V_1 \oplus \ldots \oplus V_s, V_1 \oplus \ldots \oplus V_s)$ such that: $L \neq 0$; each $c_1 \oplus \ldots \oplus c_s \in L$ has the property $\sum_i \mathrm{tr}(c_i x^{\sigma_i}) = 0$ for all $x \in G$; and L is minimal with respect to the condition $L(x^{\sigma_i} \oplus \ldots \oplus x^{\sigma_s}) = L$ for all $x \in G$. Now define $L_j = \{ c_j \mid c_1 \oplus \ldots \oplus c_s \in L$ for some c_i $(i \neq j)\}$. If $c_1 \oplus \ldots \oplus c_s \in L$ and $c_j = 0$, then the induction hypothesis shows that all $c_i = 0$. Thus $L_j \neq 0$ and so we can choose $v \in V_j$ such that the G^{σ_j}-space $v L_j$ of V_j is non-zero. By the irreducibility of σ_j, $v L_j = V_j$. Let $\rho : G \to GL(L)$ be the representation of G defined by

$$x^{\rho} : c_1 \oplus \ldots \oplus c_s \mapsto c_1 x^{\sigma_1} \oplus \ldots \oplus c_s x^{\sigma_s}.$$

Then the mapping $\theta : c_1 \oplus \ldots \oplus c_s \mapsto vc_j$ is a G-homomorphism of L onto V_j. Now σ_j is irreducible by hypothesis and ρ is irreducible by the minimality of L. Therefore by Schur's lemma (§2.6) σ_j is equivalent to ρ. But this is true for all j; since the σ_j are pairwise inequivalent, we have a contradiction. This proves the assertion for the case $s > 1$. ■

COROLLARY 2.7A. *If G is a group with exactly k classes of conjugate elements, and* $\sigma_1, \ldots, \sigma_s$ *are pairwise inequivalent irreducible representations of G over an algebraically closed field F, then* $s \leqslant k$.

Proof. If $x, y \in G$, then $\operatorname{tr}(y^{-1}xy)^{\sigma_i} = \operatorname{tr}(y^{\sigma_i})^{-1} x^{\sigma_i} y^{\sigma_i} = \operatorname{tr} x^{\sigma_i}$. Thus the value of $\operatorname{tr} x^{\sigma_i}$ only depends on the conjugacy class to which x belongs. Choose $x_1, \ldots, x_k$ as representatives of the k conjugacy classes of G, and consider the k-vectors

$$v_i = (\operatorname{tr} x_1^{\sigma_i}, \ldots, \operatorname{tr} x_k^{\sigma_i}) \quad (i = 1, \ldots, s).$$

If

$$\sum_{i=1}^{s} a_i v_i = 0 \quad \text{for some } a_i \in F,$$

then

$$\sum_{i=1}^{s} \operatorname{tr}(a_i x_j^{\sigma_i}) = 0 \quad \text{for } j = 1, \ldots, s;$$

and hence

$$\sum_{i=1}^{s} \operatorname{tr}(a_i x^{\sigma_i}) = 0 \quad \text{for all } x \in G.$$

Thus the Theorem implies that $0 = \alpha_1 = \ldots = \alpha_s$, and so the k-vectors $v_1, \ldots, v_s$ are linearly independent over F. This shows $s \leqslant k$. ■

THEOREM 2.7B. *Let F be an algebraically closed field. If G is a subgroup of $GL(n,F)$ then the following are equivalent:*

(1) *G is irreducible over F.*

(2) *$FG = M(n,F)$ (or equivalently, $\dim_F FG = n^2$).*

(3) *G is absolutely irreducible; that is, for any field extension E of F, G is irreducible as a subgroup of $GL(n, E)$.*

Proof. (3) $\Longrightarrow$ (1). Trivial.

(2) $\Longrightarrow$ (3). For, if $FG = M(n,F)$, then $EG = M(n, E)$ for any extension field E of F. But then $EG = E \cdot GL(n,E)$ and so every G-space in the underlying vector space over E is also a $GL(n, E)$-space; since $GL(n, E)$ is irreducible, G is irreducible over E. Thus G is absolutely irreducible.

(1) $\Longrightarrow$ (2). Choose elements $x_1, \ldots, x_m$ $(m \leqslant n^2)$ of G as a basis of FG over F. If $m < n^2$ and $x_s = [\xi_{ij}^{(s)}]$, then we could find a non-zero matrix $c = [\gamma_{ij}] \in M(n, F)$ such that the m equations

$$\sum_{i,\, j=1}^{n} \gamma_{ij}\, \xi_{ji}^{(s)} = 0 \quad (s = 1, \ldots, m) \tag{2.7.2}$$

are satisfied. But $\sum_{i,j} \gamma_{ij}\, \xi_{ji}^{(s)} = \operatorname{tr} cx_s$. Since each $x \in G$ is a linear combination of $x_1, \ldots, x_m$ we would have from (2.7.2) that $\operatorname{tr} cx = 0$ for all $x \in G$. By Theorem 2.7A this contradicts the irreducibility of G. Hence we conclude $m = n^2 = \dim M(n, F)$, and so $FG = M(n, F)$. ■

Exercise. Let G be an irreducible linear group of degree n over an algebraically closed field. Show that $|G : Z(G)| \geqslant n^2$. [*Hint:* Use Corollary 2.6.]

§2.8. In this section we give some immediate applications of the theorems of §2.7.

THEOREM 2.8A. *Let G be an irreducible linear group of degree n over an algebraically closed field F. If the set* $\{\mathrm{tr}\ x \mid x \in G\}$ *of trace values is finite, say of order m, then G is finite and* $|G| \leqslant m^{n^2}$

Proof. By Theorem 2.7B the vector space FG has a basis over F consisting of n^2 elements $x_i (i = 1, \ldots, n^2)$ in G. To each $y \in G$ there corresponds a list $\lambda(y) = (\mathrm{tr}\ x_1 y, \ldots, \mathrm{tr}\ x_{n^2} y)$ of n^2 trace values in G. If $\lambda(y) = \lambda(z)$, then $\mathrm{tr}\ x_i(y - z) = 0$ for each i; since each $u \in FG$ is a linear combination of the x_i, this implies $\mathrm{tr}\ u(y - z) = 0$ for all $u \in FG$. By letting u run over the different matrices with a single non-zero entry we see that this implies $y - z = 0$. Hence different elements of G are associated with different lists of trace values. Thus the order of G is at most the number of different lists, viz. m^{n^2}. ■

If an element $x \in GL(V)$ is unipotent then x corresponds to a special triangular matrix (its Jordan form) over a suitable basis for V. This observation is generalized in the following theorem of E. R. Kolchin (1948).

THEOREM 2.8B. *Let V be a vector space of dimension n over a field F, and let G be a subgroup of GL(V). If each element of G is unipotent, then over a suitable basis of V, G corresponds to a subgroup of STL(n, F).*

Remark. The converse is clearly also true. If char $F = p > 0$, then $x \in GL(V)$ is unipotent $\Longleftrightarrow$ x is a p-element. Thus the theorem shows that $STL(n, F)$ is a Sylow p-group of $GL(n, F)$ and that every p-subgroup of $GL(n, F)$ is conjugate to a subgroup of $STL(n, F)$.

Proof. We proceed by induction on n; $n = 1$ is trivial. Let $n > 1$, and first suppose G is reducible. If $W \neq 0$ or V is a G-space, then $G|W$ and $G|(V/W)$ are both groups of unipotent elements. Thus, by induction, there are bases $v_1, \ldots, v_r$ and $v_{r+1} + W, \ldots, v_n + W$ of W and V/W, respectively, over which these groups correspond to subgroups of the special triangular groups over F. Then G corresponds to a subgroup of $STL(n, F)$ over the basis $v_1, \ldots, v_n$.

It remains to show that G is always reducible when $n > 1$. First consider the case where F is algebraically closed. Since $\operatorname{tr} x = n$ for all $x \in G$, Theorem 2.8A shows that if G is irreducible then $|G| = 1$ and so $n = 1$ contrary to our supposition. Thus, if F is algebraically closed, then G is reducible and so G corresponds to a subgroup of $STL(n, F)$ over a suitable basis by the argument above. In the general case, let E be the algebraic closure of F. Then G acts as a group of unipotent elements on the vector space $V^E = V \otimes_F E$ over E. Hence for some basis of V^E, G corresponds to a subgroup of $STL(n, E)$. This implies that every product of n elements of the form $x - 1\,(x \in G)$ equals 0. Choose m such that each product of $m + 1$ elements of the form $x - 1$ $(x \in G)$ equals 0, but some product of m such elements is not zero, say $(x_1 - 1) \ldots (x_m - 1) \neq 0$. Choose $v \in V$ such that $w = v(x_1 - 1) \ldots (x_m - 1) \neq 0$. Then the 1-dimensional space W spanned by w is a G-space. Hence G is reducible (over F), and the theorem follows in the general case. ∎

Note. The last half of the proof of Theorem 2.8B shows that if G is a group of unipotent elements, then each minimal G-space has dimension 1. In particular, if G is completely reducible, $G = 1$.

The previous theorem serves to completely characterize the kernel N of the homomorphism ρ defined in Theorem 2.4.

THEOREM 2.8C. *The kernel N of the homomorphism ρ defined in Theorem 2.4 is the unique maximal normal subgroup of G which consists entirely of unipotent elements.*

Proof. The image G^{ρ} is a completely reducible linear group, and so by Theorem 2.2 each normal subgroup is completely reducible. Hence the Note following Theorem 2.8B shows that G^{ρ} has no non-trivial normal subgroup of unipotent elements. But if N_1 is a normal subgroup of unipotent elements in G, then it follows from the definition of ρ that N_1^{ρ} consists of unipotent elements; hence we conclude $N_1^{\rho} = 1$ and so $N_1 \subseteq N$. Since N consists of unipotent elements by Theorem 2.4, N has the property asserted in the theorem. ■

Let G be a completely reducible subgroup of $GL(V)$ where V is a vector space over a field F. If E is an extension field of F and $V^E = V \otimes_F E$, then G may be considered as a subgroup of $GL_E(V^E)$. It is not known, in general, whether G is necessarily completely reducible over E (but see §2.10). However, G is isomorphic to a completely reducible subgroup of $GL(V^E)$. Specifically, the following is true.

COROLLARY 2.8C. *With the notation above, if we consider G as a subgroup of $GL(V^E)$, then any representation $\rho : G \to GL(V^E)$ of the kind described in Theorem 2.4 is faithful.*

Proof. Because G is a completely reducible subgroup of $GL(V)$, it follows from the Theorem that the largest normal subgroup of G consisting of unipotent elements is 1. Hence $\ker \rho = 1$. ■

Exercises. 1. Let G be a linear group of degree n. If, for some integer $m > 0$, x^m is unipotent for each $x \in G$, show that G has a normal subgroup H consisting of unipotent elements such that $|G : H| \leqslant m^{n^3}$.

2. Let V be a vector space of dimension $n > 1$ over a field of characteristic $p > 0$. Show that each p-subgroup $\neq 1$ of $GL(V)$ is reducible.

§2.9. The theorem of this section (at least in the case of characteristic 0) is due to W. Burnside.

THEOREM 2.9. *Let G be a subgroup of $GL(V)$ where V is a vector space of dimension n over a field F. Suppose G has finite exponent e, that is e is an integer > 0 and $x^e = 1$ for all $x \in G$. Then:*

(1) *If* char $F = 0$, *then G is finite and* $|G| \leqslant e^{n^3}$.

(2) *If* char $F = p > 0$, *then G has a normal nilpotent p-subgroup N (consisting of unipotent elements) such that* $|G:N| \leqslant e^{n^3}$; *in particular, $N = 1$ if either $p \nmid n$ or G is completely reducible.*

Proof. We first apply Theorem 2.4 to reduce to the completely reducible case. If char $F = 0$, then $\ker \rho = 1$ in Theorem 2.4 because G is periodic. Similarly, if char $F = p > 0$, then $\ker \rho$ is a p-group N, and $N = 1$ if either $p \nmid e$ or G is completely reducible. It follows from Corollary 2.8C that without loss in generality we may suppose that F is algebraically closed.

Thus it remains to show that if G is a completely reducible linear group of degree n over an algebraically closed field F, and G has exponent e, then $|G| \leqslant e^{n^3}$. An obvious induction shows that we need only consider the case where G is irreducible. However, there are at most e different $\xi \in F$ such that $\xi^e = 1$, and so there are at most e^n possible trace values for $x \in G$, (since the eigenvalues of x are all eth roots of 1). By Theorem 2.8A we conclude $|G| \leqslant e^{n^3}$ as required. ■

Exercise. Let G be a linear group of degree *n*. If G has only a finite number of conjugacy classes prove that G is finite.

§2.10. In this section, we examine what happens to irreducibility when we extend the ground field. We begin with a simple consequence of Burnside's theorem (Theorem 2.7B).

THEOREM 2.10A. *Let G be an irreducible subgroup of GL(n, F) over a field F. Then, either G is absolutely irreducible, or G is reducible over some finite field extension E of F (that is, G is reducible when considered as a subgroup of GL(n, E)).*

Proof. Suppose G is not absolutely irreducible. Then, by Theorem 2.7B, G is reducible over the algebraic closure $\overline{F}$ of F. Thus there exists $c \in GL(n, \overline{F})$ such that for some integer r, $1 \leqslant r < n$,

$$c^{-1} x c = \begin{bmatrix} x_1 & 0 \\ x_3 & x_2 \end{bmatrix} \quad \text{for all } x \in G,$$

where x_1, x_2 and x_3 are blocks of dimensions $r \times r$, $(n-r) \times (n-r)$ and $(n-r) \times r$, respectively. Define E as the field generated over F by the entries in c. Then E is generated by a finite number of elements algebraic over F, so $[E:F] < \infty$. But $c \in GL(n, E)$ so G is reducible over E. ■

Note. If F is a perfect field (in particular, if F is finite or char $F = 0$), then each finite extension of F is separable. Hence in this case we can even choose the field E in Theorem 2.10A to be a Galois extension of F, that is, a normal, separable extension.

THEOREM 2.10B. *Let G be an irreducible subgroup of GL(n, F) and let E be a finite Galois extension of the field F. Then, for some*

$c \in GL(n, E)$, *some integer* $m \geqslant 1$, *and some distinct automorphisms* $\alpha_1 = 1, \alpha_2, \ldots, \alpha_m \in \mathrm{Gal}\,(E/F)$ *(the Galois group of* E *over* F*), we have*

$$c\,x\,c^{-1} = \mathrm{diag}\,(x_1^{\alpha_1}, x_1^{\alpha_2}, \ldots, x_1^{\alpha_m}) \quad \textit{for all } x \in G, \tag{2.10.1}$$

where $x \leftrightarrow x_1$ $(x \in G)$ *is an irreducible matrix representation of* G *of degree* $d = n/m$ *over* E *and* $x_1^{\alpha_s} \overset{\text{def}}{=} [\xi_{ij}^{\alpha_s}]$ *when* $x_1 = [\xi_{ij}]$. *In particular:*

(1) G *is completely reducible over* E *and all its irreducible components have the same degree* d.

(2) *The representation* $x \leftrightarrow x_1$ *of* G *over* E *is faithful because* $x_1 = 1$ *implies* $x_1^{\alpha_s} = 1$ *for all* s.

Our proof of this theorem requires the following result.

LEMMA 2.10. *Let* E *be a finite Galois extension of a field* F *and let* $\alpha_1, \ldots, \alpha_m$ *be distinct elements of* $\mathrm{Gal}(E/F)$. *If* $\lambda_1, \ldots, \lambda_m$ *are elements of* F *and are not all zero then, for some* $\xi \in E$, $\lambda_1 \xi^{\alpha_1} + \ldots + \lambda_m \xi^{\alpha_m} \neq 0$.

Proof. Let E^* be the multiplicative group of non-zero elements of E. Then we have first degree (and hence absolutely irreducible) representations of E^* over E given by $\xi \leftrightarrow \xi^{\alpha_s}$ for each s. Since these representations are pairwise inequivalent, Theorem 2.7A shows that

$$\sum_{s=1}^{m} \lambda_s \xi^{\alpha_s} = 0 \text{ (for all } \xi \in E^*) \Longrightarrow \text{all } \lambda_s = 0. \;\blacksquare$$

Proof of Theorem 2.10B. Let V be the underlying vector space of $GL(n, F)$ and V^E the underlying vector space of $GL(n, E)$. Choose $b \in GL(n, E)$ such that for some integer d, $1 \leqslant d \leqslant n$,

$$bxb^{-1} = \begin{bmatrix} x_1 & 0 \\ x_3 & x_2 \end{bmatrix} \quad \text{for all } x \in G \qquad (2.10.2)$$

with x_1, x_2 and x_3 blocks of dimensions $d \times d$, $(n-d) \times (n-d)$ and $(n-d) \times d$, respectively, and where $x \mapsto x_1$ is an irreducible matrix representation of G over E. Let W be the subspace of V^E consisting of all vectors whose last $n - d$ entries are 0. Then, for each $\alpha \in \mathrm{Gal}(E/F)$, (2.10.2) shows

$$Wb^\alpha x = W(bxb^{-1})^\alpha b^\alpha = Wb^\alpha \quad \text{for all } x \in G,$$

because $x \in GL(n, F)$ implies $x^\alpha = x$. Thus we conclude that Wb^α is an irreducible G-space in V^E for each $\alpha \in \mathrm{Gal}(E/F)$. Now choose $v \in W$ such that some entry in the vector vb has the property $\sum_\alpha \xi^\alpha \neq 0$ (α runs over $\mathrm{Gal}(E/F)$); this can be done by multiplying a non-zero vector by a suitable scalar and using Lemma 2.10. Then $\sum_\alpha (vb)^\alpha$ is a non-zero vector in $W_1 = (\sum_\alpha Wb^\alpha) \cap V$. But W_1 is a G-space, and so $W_1 = V$ because G is irreducible on V. Thus $V^E = V \otimes_F E \subseteq \sum_\alpha Wb^\alpha$ and so $V^E = \sum_\alpha Wb^\alpha$. Since each Wb^α is a minimal G-space in V^E it follows that $V^E = \bigoplus_{i=1}^m Wb^{\alpha_i}$ for some distinct $\alpha_1 = 1, \ldots, \alpha_m$ in $\mathrm{Gal}(E/F)$ (compare the proof of Theorem 2.2). Finally, choose vectors $v_j \in V$ such that $v_j (j = i\text{-}1)d + 1, \ldots, id)$ is a basis for Wb^{α_i} for $i = 1, \ldots, m$. It is now readily verified that the matrix $c \in GL(n, E)$ whose jth row is v_j ($j = 1, \ldots, n$) has the property asserted in the theorem. ∎

Exercises. 1. Let G be a finite irreducible linear group of degree n over a perfect field F. If G is abelian, show that

(a) G is cyclic (with order prime to char F if char $F > 0$).

(b) The minimal polynomial of any generator x of G has degree n.

2. In general the representations $x \mapsto x_1^{\alpha_s}$ of G defined in Theorem 2.10B are not all different. However, prove that:

(a) If all these representations are equivalent, then G is irreducible over E (so $m = 1$).

(b) The multiplicity to which a given irreducible component occurs is the same for each of the inequivalent components.

3. Give a proof of Lemma 2.10 without using Theorem 2.7A.

§2.11 *Notes and References.*

The theorems of Clifford and Maschke are basic results in the theory of representations and characters. The analysis of Theorem 2.2 is extended further in the original paper

Clifford, A. H. Representations induced in an invariant subgroup, *Ann. Math.* **38**, 533–550 (1937).

E. Dade has recently extended these results further in an unpublished manuscript (University of Illinois). The generalization of Maschke's theorem which we give seems to have been discovered independently by several authors.

Theorem 2.4 appears explicitly in:

Dixon, J. D. The solvable length of a solvable linear group, *Math. Z.* **107**, 151–158 (1968).

but the idea was used much earlier.

The theorems of §2.7 can be deduced as corollaries to the Wedderburn structure theorems on semisimple rings. For example, see §27 of the book of Curtis and Reiner (§0.4). I believe that the approach here was first used by H. Weyl.

Theorems 2.8A and 2.9 (in the case of the complex field) and the Exercise of §2.9 were proved in

Burnside, W. On criteria for the finiteness of the order of a group of linear substitutions, *Proc. Lond. Math. Soc.* (2) **3**, 435–440 (1905).

Theorem 2.8B appears in

Kolchin, E. R. Algebraic matric groups and the Picard–Vissiot theory of homogeneous linear differential equations, *Ann. Math.* (2) **49**, 1–42 (1948).

Our proof of this theorem follows

Kaplansky, I. Notes on ring theory (mimeographed notes), University of Chicago (1965).

For an alternative exposition of the results in §2.10 see §69 of the book of Curtis and Reiner. Lemma 2.10 is due to Artin and has numerous applications; see page 209 of Lang's book (§0.4).

CHAPTER 3

Changing the Ground Ring

§3.1. A very useful technique in analysing the structure of a linear group is that of 'changing the ground ring'. Strictly, it is a device for getting a new representation of the group from the given one. In the case of finite groups we are usually looking for faithful representations over some more easily handled ground ring such as the complex field. In the case of infinite groups we can use the same technique to get proper homomorphic images and hence get information on the normal structure of the group (see Chapter 10).

Definitions and examples are given in §3.2. In §3.3 we prepare for the proof of Theorem 3.4A which has many important applications in what follows. The principal theorem of the next two sections is Theorem 3.6B which shows how we can change from a field of characteristic 0 to one of characteristic $p > 0$, and Theorem 3.7 strengthens this result. Perhaps the main result of this Chapter, at least as far as finite groups goes, is Theorem 3.8 which shows how, for a finite group G, we can obtain a faithful representation of degree n over the complex field from a faithful representation of degree n over a field of characteristic $p > 0$, provided $p \nmid |G|$. Section 3.9 generalizes this result.

§3.2 Let $\varphi : R \to S$ be a homomorphism from a ring R into a ring S (our convention is that all rings have unity 1 and that a homomorphism maps unity onto unity). For each integer $n \geqslant 1$, φ induces a homomorphism $M(n, R) \to M(n, S)$ defined by $[\xi_{ij}] \mapsto [\xi_{ij}^{\varphi}]$. By restriction, this latter mapping gives a group homomorphism $\varphi^* : G \to GL(n, S)$ for any subgroup G of $GL(n, R)$; indeed, if $[\xi_{ij}] \in G$ has inverse $[\eta_{ij}]$, then $[\eta_{ij}^{\varphi}]$ is the inverse of $[\xi_{ij}^{\varphi}]$, so $[\xi_{ij}^{\varphi}] \in GL(n, S)$. We call φ^* the *homomorphism induced on G by* φ. Note that φ^* is injective if φ is.

Examples. 1. Let R be a subring of S, and let $\varphi : R \to S$ be the embedding. The induced homomorphism is used so often we usually do not bother to point it out.

2. Let I be an ideal in a ring R and let $\varphi : R \to R/I$ be the natural homomorphism. We shall refer to both φ and the induced homomorphism φ^* on any subgroup of $GL(n, R)$ as the mapping (mod I).

3. If E is a field extension of degree d over a field F, then E is a vector space of dimension d over F and there is an isomorphism $E \to \overline{E} \subseteq \mathrm{Hom}_F(E, E)$ defined by $x \mapsto \overline{x}$ where $\overline{x} : u \mapsto ux\,(u,\ ux \in E)$. By choosing any basis for E over F we can thus obtain an injective homomorphism $\varphi : E \to M(d, F)$. Then, for each subgroup G of $GL(n, E)$ there is the induced isomorphism $\varphi^* : G \to \overline{G} \subseteq GL(n, M(d, F))$. But the latter is a group of block matrices with $d \times d$ blocks over F. Hence we have $GL(n, M(d, F)) = GL(nd, F)$ and φ^* is an isomorphism of G onto a subgroup of $GL(nd, F)$.

Exercise. Let G be a subgroup of $GL(n, \mathbf{C})$ ($\mathbf{C}$ is the field of complex numbers). If there are real constants $c_1 > c_2 > 0$ such that the absolute value of each non-zero entry in each matrix in G lies between c_1 and c_2, show that G is finite. [*Hint:* Use Example 3 to reduce to the case of a group over the real numbers.]

§3.3 This section deals with two general results on the existence of common zeros of a family of polynomials. The first theorem will be used in Chapter 10, but the second theorem will be applied immediately in the next section.

Let F be a field and let $X = (X_\lambda)_\lambda$ be a family of indeterminates. We write $F[X]$ to denote the ring of all polynomials $f(X)$ over F in the indeterminates X_λ. Although X may be an infinite family, of course $f(X)$ will only involve a finite number of X_λ. A family Δ of polynomials in X over F has a *common zero* $\xi = (\xi_\lambda)_\lambda$ in extension field E of F if all $\xi_\lambda \in E$ and the substitution $X \mapsto \xi$ sends $f(X) \mapsto f(\xi) = 0$ for all $f(X) \in \Delta$.

THEOREM 3.3A. *Let $\Delta \subseteq F[X]$. Then Δ has a common zero in some extension field of $F \Longleftrightarrow$ there do not exist $f_1(X), \ldots, f_s(X) \in \Delta$ and $g_1(X), \ldots, g_s(X) \in F[X]$ such that*

$$\sum_{i=1}^{s} f_i(X)\, g_i(X) = 1. \tag{3.3.1}$$

Proof. If (3.3.1) holds then $\{f_1(X), \ldots, f_s(X)\}$ obviously has no common zero and so Δ has no common zero.

Conversely, suppose no equation of the form (3.3.1) holds. Let I be the ideal generated by Δ in $F[X]$. Our hypothesis implies that $1 \notin I$, so by using Zorn's lemma we can show that there exists an ideal M of $F[X]$ which is maximal with respect to the conditions $1 \notin M$ and $I \subseteq M$. It is readily verified that M is a maximal ideal of $F[X]$; and therefore $E = F[X]/M$ is a field. The natural mapping $F[X] \to F[X]/M$ embeds F into E and we shall identify F with its image; then E is an extension field of F. Furthermore, let ξ_λ be the image of X_λ under this mapping. Then each $\xi_\lambda \in E$ and $\xi = (\xi_\lambda)_\lambda$ is a common zero of Δ because $f(X) \in \Delta \subseteq I \subseteq M$ implies $f(\xi) = 0$. ■

Note. The extension field E is actually the ring extension $F[\xi]$, that is the set of all polynomials in ξ over F.

We shall now consider the case of a finite number of indeterminates a little more carefully. We need the following lemma.

LEMMA 3.3. *Let R be a subring of a field E_1 and let E be a finite extension field of E_1. If E is a finitely generated ring extension of R, say $E = R[\xi_1, \ldots, \xi_n]$, then E_1 is a finitely generated ring extension of R.*

Proof. Since E is a finite dimensional vector space over E_1, $E = E_1\omega_1 \oplus \ldots \oplus E_1\omega_m$ for some basis $\omega_1, \ldots, \omega_m$. Define $a_{ijk}, \beta_{st} \in E_1$ by

$$\omega_i\omega_j = \sum_k a_{ijk}\,\omega_k \quad (i, j = 1, \ldots, m)$$

$$\xi_s = \sum_t \beta_{st}\,\omega_t \quad (s = 1, \ldots, n)$$

and let E_0 be the finitely generated ring extension of R generated by the a_{ijk}, β_{st}. We conclude the proof by showing $E_0 = E_1$. Certainly $E_0 \subseteq E_1$. On the other hand $E_0\omega_1 + \ldots + E_0\omega_m$ is a subring of E containing R and all ξ_s; hence $E = E_0\omega_1 + \ldots + E_0\omega_m$. Since $E = E_1\omega_1 \oplus \ldots \oplus E_1\omega_m$ and $E_0 \subseteq E_1$, we must have $E_1\omega_i \subseteq E_0\omega_i$ for each i. Hence $E_0 = E_1$. ■

THEOREM 3.3B. *Let Δ be a family of polynomials in a finite family $X = (X_i)_{i=1}^n$ of indeterminates over a field F. If Δ has a common zero in some extension field of F, then there is an extension field E of finite degree over F in which Δ has a common zero.*

Proof. We shall show that the field E defined in Theorem 3.3A has finite degree over F. We noted at the end of the proof of that theorem that $E = F[\xi_1, \ldots, \xi_n]$, so $[E:F] < \infty$ unless some ξ_i is not algebraic over F.

Suppose $[E:F] = \infty$. Without loss in generality we may suppose $\xi_1, \ldots, \xi_r$ $(r \geqslant 1)$ are algebraically independent over F, and the remaining ξ_i are algebraically dependent on these. Thus if we put $E_1 = F(\xi_1, \ldots, \xi_r)$, then E is a finitely generated algebraic extension of E_1 and so $[E:E_1] < \infty$. From Lemma 3.3 there exist $\eta_i \in E_1$ such that $E_1 = F[\eta_1, \ldots, \eta_m]$. Each η_i is a quotient of elements in $F[\xi_1, \ldots, \xi_r]$, and since $F[\xi_1, \ldots, \xi_r]$ $(\simeq F[X_1, \ldots, X_r])$ has an infinte number of different irreducible elements, we may choose one, say λ, such that λ does not divide the denominator of any η_i. Then $\lambda \in E_1$, but $1/\lambda \notin F[\eta_1, \ldots, \eta_m] = E_1$-contrary to the fact that E_1 is a field. Hence we conclude that $[E:F] < \infty$. ■

§3.4 This section gives some applications of Theorem 3.3B to the theory of representations of finite groups. The basic idea (due to Mal'cev) is that a matrix representation of a given degree of a group G is simply a common zero of a family of polynomials which can be written down from a knowledge of the group table.

LEMMA 3.4. *Let G be a group and let ρ and σ be representations of the same degree n over an algebraically closed field F. If σ is irreducible and* $\operatorname{tr} x^\sigma = \operatorname{tr} x^\rho$ *for all $x \in G$, then σ is equivalent to ρ.*

Proof. By suitable choices for the bases of the underlying vector spaces we may suppose both ρ and σ are matrix representations over F and that the matrices in G^ρ have the form

$$x^{\rho} = \begin{bmatrix} x^{\rho_1} & & & & 0 \\ & x^{\rho_2} & & & \\ & & \cdot & & \\ & * & & \cdot & \\ & & & & x^{\rho_s} \end{bmatrix} \quad \text{for all } x \in G$$

where the mappings $x \mapsto x^{\rho_i}$ are irreducible representations of G for $i = 1, \ldots, s$. By hypothesis $0 = \operatorname{tr} x^{\rho} - \operatorname{tr} x^{\sigma} = \sum_{i=1}^{s} \operatorname{tr} x^{\rho_i} + \operatorname{tr} (-1)x^{\sigma}$ for all $x \in G$. On collecting together the equivalent representations, we see from Theorem 2.7A that at least one ρ_j is equivalent to σ. Since $\deg \sigma = n = \Sigma \deg \rho_i$, we conclude $s = 1$ and $\rho = \rho_j$ is equivalent to σ. ■

Recall from Theorem 2.7B that an irreducible representation over an algebraically closed field is absolutely irreducible.

THEOREM 3.4A. *Let G be a finite group and let $\sigma : G \to GL(n, F)$ be an irreducible matrix representation over an algebraically closed field F. Let F_1 be a subfield of F containing the values of the character χ afforded by σ. Then there is a finite extension field E of F_1 in F and a representation $\rho : G \to GL(n, E)$ such that ρ is equivalent to σ (when we consider ρ as a representation over F).*

Proof. We introduce the finite family $X = (X_{ijx})$ of indeterminate $(i, j = 1, \ldots, n;\ x \in G)$, and let Δ be the family of all polynomials in $F_1[X]$ of the following types:

(a) entries in the $n \times n$ matrix

$$[X_{ijx}][X_{ijy}] - [X_{ij(xy)}] \quad \text{for } x, y \in G \tag{3.4.1}$$

(b)

$$\det[X_{ij1}] - 1 \tag{3.4.2}$$

(c) $$\sum_{i=1}^{n} X_{iix} - \chi(x) \quad \text{for } x \in G \tag{3.4.3}$$

Then Δ has a common zero in F given by the substitution $[X_{ijx}] \mapsto x^\sigma$ $(x \in G)$. Thus by Theorem 3.3B there is a finite extension E of F_1 in which Δ has a common zero; since F contains the algebraic closure of F_1 we may take $E \subseteq F$. This means that for each $x \in G$ we have a matrix $x^\rho \in M(n, E)$ such that the substitution $[X_{ijx}] \mapsto x^\rho$ $(x \in G)$ sends each polynomial in Δ to 0. From (3.4.1) and (3.4.2) we have $(xy)^\rho = x^\rho y^\rho$ and $\det(x^\rho (x^{-1})^\rho) = 1$ (so x^ρ is non-singular) for all $x, y \in G$. Thus we have found a representation $\rho : G \to GL(n, E)$ and (3.4.3) shows that $\operatorname{tr} x^\rho = \chi(x) = \operatorname{tr} x^\sigma$ for all $x \in G$. Since σ is irreducible, Lemma 3.4 shows that ρ is equivalent to σ. ∎

COROLLARY 3.4A. *There exists a finite extension E in F of the prime subfield of F such that each irreducible representation of G over F is equivalent to a representation over E. Moreover, if char $F = 0$ or char $F \nmid |G|$, then every representation of G over F is equivalent to a representation over E.*

Proof. By Corollary 2.6A there are only a finite number of pairwise inequivalent irreducible representations of G over F. Thus the first half of our Corollary follows from the Theorem by an obvious induction. In the case that char $F = 0$ or char $F \nmid |G|$, each representation of G over F is completely reducible by Theorem 2.3, and so by taking the irreducible components separately it follows that each representation over F is equivalent to one over E. ∎

We reformulate Theorem 3.4A in terms of matrix groups.

THEOREM 3.4B. *Let G be a finite irreducible subgroup of $GL(n, F)$ where F is an algebraically closed field. Then there is a finite extension field E in F of the prime subfield of F such that G is conjugate in $GL(n, F)$ to a subgroup of $GL(n, E) \subseteq GL(n, F)$.*

Note. In the case char $F = 0$, the prime subfield of F is isomorphic to the rational field $\mathbf{Q}$. Each finite extension of $\mathbf{Q}$ is isomorphic to a subfield of the complex field $\mathbf{C}$ because the latter is algebraically closed. Hence Corollary 3.4A shows that a finite linear group over a field of characteristic 0 is isomorphic to a linear group of the same degree over $\mathbf{C}$.

Exercises. 1. Show that the condition in Lemma 3.4 that $\deg \rho = \deg \sigma$ may be dropped if char $F = 0$; but the condition is necessary when char $F > 0$.

2. Let F be the algebraic closure of a field with two elements, and let $E = F(\alpha)$ be a transcendental extension of F. Show that the representation of the group $G = \langle a, b | a^2 = b^2 = (ab)^2 = 1 \rangle$ of order 4 defined by

$$a \leftrightarrow \begin{matrix} 1 & 0 \\ 1 & 1 \end{matrix} \quad \text{and} \quad b \leftrightarrow \begin{matrix} 1 & 0 \\ \alpha & 1 \end{matrix}$$

is *not* equivalent to any representation over F. (Compare with Corollary 3.4A)

§3.5. An early theorem of W. Burnside (1911) showed that each finite subgroup of $GL(n, \mathbf{Q})$ is conjugate to a subgroup of $GL(n, \mathbf{Z}) \subseteq GL(n, \mathbf{Q})$. This was later generalized as follows.

THEOREM 3.5. *Let D be a principal ideal domain (PID) and let F be its field of quotients. Let G be a finite subgroup of $GL(V)$*

where V is a vector space of dimension n over F. Then, over a suitable basis of V, G corresponds to a subgroup of $GL(n, D) \subseteq GL(n, F)$.

Proof. Choose any basis $v_1, \ldots, v_n$ of V and define

$$M = \{v \in V | vx \in Dv_1 + \ldots + Dv_n \text{ for all } x \in G\} \subseteq Dv_1 + \ldots + Dv_n.$$

Observe that M is a D-submodule of V and that $Mx \subseteq M$ for all $x \in G$. Because F is the field of quotients of D and G is finite, for each $w \in V$ there exists $\delta \neq 0$ in D such that $\delta w \in M$. Hence all the sets (for $j = 1, \ldots, n$) $I_j = \{a_j \in D |$ there exist $a_i \in D$ $(i < j)$ such that $\sum_{i=1}^{j} a_i v_i \in M\}$ contain some non-zero elements. Now each I_j is an ideal in D so for some $\beta_{jj} \neq 0$ in $D: I_j = \beta_{jj}D$. We choose $u_j \in M$ such that $u_j = \sum_{i=1}^{j} \beta_{ji}v_i$ for some $\beta_{ji}, \ldots, \beta_{jj-1} \in D$. Since the β_{jj} are non-zero, $u_1, \ldots u_n$ is a basis for V. Moreover, if $v = \sum_{i=1}^{n} a_i v_i \in M$, and a_k is the last non-zero coefficient, then $a_k = \gamma_k \beta_{kk}$ for some $\gamma_k \in D$ by the choice of β_{kk}; hence $v - \gamma_k u_k \in Dv_1 + \ldots + Dv_{k-1}$. Therefore an induction argument shows that $v = \sum_{i=1}^{n} \gamma_i u_i$ for some $\gamma_i \in D$. Hence we conclude $M = Du_1 \oplus \ldots \oplus Du_n$. It is now clear that G corresponds to a subgroup of $GL(n, D)$ over the basis $u_1, \ldots u_n$. ■

Exercise. Let E be an extension field of a field F and let D_E and D_F denote subrings of E such that E and F are fields of quotients of D_E and D_F, respectively. If G is a subgroup of $GL(n, D_F)$, and G is reducible over E, show that G is reducible as a subgroup of $GL(n, D_E)$.

§3.6. To apply Theorem 3.5 we need to construct suitable PIDs. One way of doing this is the following.

Let $F \subseteq \mathbf{C}$ be a finite extension field of the rational field $\mathbf{Q}$, and let D be the ring of algebraic integers in F. For a given rational prime p we choose a maximal (prime) ideal I_0 in D with $p \in I_0$. Now a classical theorem of Dedekind states that each non-zero ideal in D may be factored into a product of prime ideals in D in exactly one way (see H. Pollard, *Theory of Algebraic Numbers*, Wiley). Thus we can define the *valuation* $\nu = \mathrm{v}_{I_0}$ on D by defining $\nu(a)$ as the exponent of I_0 occurring in the prime decomposition of the ideal aD if $a \neq 0$, and putting $\nu(0) = \infty$. It is readily verified that

$$\nu(a) = \nu(-a) \tag{3.6.1}$$

$$\nu(a\beta) = \nu(a) + \nu(\beta) \tag{3.6.2}$$

and

$$\nu(a + \beta) \geqslant \min\{\nu(a), \nu(\beta)\} \tag{3.6.3}$$

for all $a, \beta \in D$ with the usual conventions about ∞. Since each element of F has the form a/β ($a, \beta \in D, \beta \neq 0$), we can extend the definition of ν to F by writing $\nu(a/\beta) = \nu(a) - \nu(\beta)$; the condition (3.6.2) ensures that $\nu(\gamma/\delta) = \nu(a/\beta)$ when $\gamma/\delta = a/\beta$. It is now straightforward to verify that (3.6.1)–(3.6.3) hold for all $a, \beta \in F$. We call ν a *valuation* on F *at* p.

The *valuation ring* $D_\nu \overset{\text{def}}{=\!=} \{a \in F \mid \nu(a) \geqslant 0\}$ has the following properties.

(V1). D_ν is a subring of the field F and hence an integral domain; $a \in D_\nu$ is a unit $\Longleftrightarrow \nu(a) = 0$.

(V2). D_ν is a PID. Indeed, if $J \neq 0$ is an ideal in D_ν, then choose $a \in J$ such that the positive integer $\nu(a)$ is as small as possible. Then $\nu(\beta/a) \geqslant 0$ for all $\beta \in J$, so $\beta = a \cdot \beta/a \in aD_\nu$;

hence $J \subseteq aD_\nu$. Trivially $aD_\nu \subseteq J$ and so $J = aD_\nu$.

(V3). $M_\nu = \{a \in D_\nu | \nu(a) > 0\}$ is a proper ideal of D_ν. Since no proper ideal can contain a unit, and M_ν contains all non-units of D_ν by (V1), we conclude M_ν is the unique maximal ideal of D_ν. Since $p \in M_\nu$, $M_\nu \cap \mathbf{Z} = p\mathbf{Z}$. Thus the field $F_\nu = D_\nu/M_\nu$ has characteristic p. It is readily verified that F_ν is algebraic over its prime subfield because each element of D_ν is algebraic over $\mathbf{Q}$ (in fact, F_ν is finite – see Exercise 1).

(V4). $D_\nu \supseteq D$ and so F is the field of quotients of D_ν.

The next theorem now follows immediately from Theorem 3.5.

THEOREM 3.6A. *Let $F \subseteq \mathbf{C}$ be an extension field of finite degree over the rational field $\mathbf{Q}$, and let G be a finite subgroup of $GL(n, F)$. If D is the ring of algebraic integers in F, and D_ν is the valuation ring corresponding to some prime ideal I_0 of D, then G is conjugate in $GL(n, F)$ to a subgroup of $GL(n, D_\nu) \subseteq GL(n, F)$.* ■

The interest in Theorem 3.6A is indicated by the next result.

THEOREM 3.6B. *Using the notation above, let G be a subgroup of $GL(n, D_\nu)$. Then the mapping* (mod M_ν) (see §3.2) *maps G homomorphically into $GL(n, F_\nu)$ where $F_\nu = D_\nu/M_\nu$ is a field of characteristic p. If G is periodic, then the kernel of this homomorphism is a (normal) p-subgroup of G. In particular, if* 1 *is the only normal p-subgroup of G, then the mapping* (mod M_ν) *is an isomorphism of G onto a subgroup of $GL(n, F_\nu)$.*

Proof. The first part follows from §3.2 and the observations at the beginning of this section. It remains to show that when G is periodic and $x \in G$, then $x \equiv 1 \pmod{M_\nu}$ implies x is a p-element. To do this we first extend the definition of valuation to the set $M(n, F)$ by writing $\nu(a) = \min_{i,j} \nu(a_{ij})$ when $a = [a_{ij}] \in M(n, F)$.

From (3.6.1)–(3.6.3) we see that for all a, $b \in M(n, F)$: $\nu(a) = \nu(-a)$, $\nu(ab) \geqslant \nu(a) + \nu(b)$ and $\nu(a + b) \geqslant \min\{\nu(a), \nu(b)\}$. Now suppose $x \in G$ has order $h > 1$ and $x \equiv 1 \pmod{M_\nu}$. Then $x = 1 + y$ where $1 \leqslant \nu(y) < \infty$. But $(1 + y)^h = x^h = 1$ so the binomial theorem shows that $hy = -\sum_{i=2}^{h} \binom{h}{i} y^i$. Therefore $\nu(hy) \geqslant \nu(y^2) \geqslant 2\nu(y)$, and so $\nu(h) \geqslant \nu(y) \geqslant 1$. Hence the rational integer $h \in I_0$. Since $p \in I_0$ and $1 \notin I_0$ we conclude $p|h$. Then $x^p \equiv 1 \pmod{M_\nu}$ and x^p has order h/p, so induction on the order of x shows that x^p, and hence x, is a p-element. ∎

Exercises. 1. Show that $F = D\ /M \cong D/I$. Hence F_y has at most p^d elements where $d = [F:\mathbf{Q}] = [D:\mathbf{Z}]$ (see §19 of Curtis and Reiner's book).

2. If $F = \mathbf{Q}$, show that $D_\nu = \mathbf{Z}$ and $M_\nu = p\mathbf{Z}$.

3. Let G be a subgroup of $GL(2, \mathbf{Z})$ generated by

$$\begin{bmatrix} 0 & 1 \\ 1 & 0 \end{bmatrix} \quad \text{and} \quad \begin{bmatrix} -1 & 0 \\ 0 & 1 \end{bmatrix}.$$

Show that the homomorphism (mod 2) has a non-trivial kernel.

4. Let G be a finite subgroup of $GL(n, \mathbf{Z})$. Show that the homomorphism (mod p) for a rational prime p has kernel 1 if p is odd and kernel of order 1 or 2 if $p = 2$.

5. Let p be a prime > 3, and let $\overline{G} = STL(3, \mathbf{Z}/p\mathbf{Z})$. Show that $\overline{G}$ is a non-abelian group of order p^3. On the other hand show that 1 is the only p-subgroup of $GL(3, \mathbf{Z})$. Thus $\overline{G}$ is not the image (mod p) of any finite subgroup of $GL(3, \mathbf{Z})$. More generally, a finite p-subgroup of $GL(3, D)$ where D is an integral domain of characteristic 0 is always abelian (see Corollary 5.2), so $\overline{G}$ cannot

arise as an image of any group by the process described in Theorem 3.6B.

§3.7. We now strengthen the conclusion of Theorem 3.6B in the case that the group has order not divisible by the prime involved. It will follow that in this case we may invert the procedure described in Theorem 3.6B (see §3.8).

We need an orthogonality relation.

LEMMA 3.7. *Let G be a finite irreducible subgroup of order g in $GL(n, \mathbf{C})$, and write $x = [\xi_{ij}(x)]$ for each $x \in G$. Then*

$$\sum_{x \in G} \xi_{ij}(x)\, \xi_{k\ell}(x^{-1}) = g\, \delta_{i\ell}\, \delta_{jk}/n. \tag{3.7.1}$$

Proof. Let e_{jk} be the $n \times n$ matrix with (j, k)th entry 1 and all other entries 0. Then the left-hand side of (3.7.1) is the (i, ℓ)th entry in the matrix $z = \sum_{x \in G} x\, e_{jk}\, x^{-1}$. Since $yz = \sum_{x \in G} yx\, e_{jk}\, (yx)^{-1} y = zy$ for all $y \in G$, Corollary 2.6 shows that z is a scalar, say $\zeta 1$ $(\zeta \in \mathbf{C})$. Then $n\zeta = \operatorname{tr} z = \sum_{x \in G} \operatorname{tr} e_{jk} = g\, \delta_{jk}$. Thus the left-hand side of (3.7.1) equals the (i,ℓ)th entry in $z = \zeta 1$ where $\zeta = g\, \delta_{jk}/n$, and hence equals the right-hand side of (3.7.1). ∎

THEOREM 3.7. *Let $F \subseteq \mathbf{C}$ be an extension field of finite degree over $\mathbf{Q}$. Let p be a prime, ν a valuation of F at p, and D_ν the valuation ring with maximal ideal M_ν (see §3.6). If G is a finite absolutely irreducible subgroup of $GL(n, D_\nu)$ and $p \nmid g = |G|$, then the image of $G(\text{mod } M_\nu)$ is absolutely irreducible subgroup of $GL(n, F_\nu)$ where $F_\nu = D_\nu/M_\nu$.*

Proof. With the notation of Lemma 3.7, consider the $n^2 \times g$ matrix

$$c = \begin{bmatrix} \xi_{11}(x_1) & \xi_{11}(x_2) & \cdots & \xi_{11}(x_g) \\ \xi_{12}(x_1) & \xi_{12}(x_2) & \cdots & \xi_{12}(x_g) \\ \vdots & & & \vdots \\ \xi_{nn}(x_1) & \xi_{nn}(x_2) & \cdots & \xi_{nn}(x_g) \end{bmatrix},$$

where $x_1, \ldots, x_g$ are the elements of G. Let c_0 be the matrix obtained from c by interchanging the ith and jth columns whenever $x_i = x_j^{-1}$ and interchanging the $(k\ell)$th row with the (ℓk)th row for all $k, \ell = 1, \ldots, n$. If c_0^T is the transpose of c_0, then $cc_0^T = (g/n)1_{n^2}$ by Lemma 3.7. Since both c and c_0 have all entries in D_ν, $g/n \in D_\nu$. Using bars to denote the image (mod M_ν) we obtain $\overline{cc_0^T} = (\overline{g/n})1_{n^2}$. Since $p \nmid g$, $\overline{g/n} \neq 0$ (see (V3) of §3.6) and so rank $\bar{c} \geqslant$ rank $\bar{c}\,\bar{c}_0^T = n^2$. Thus $\bar{c}$ has n^2 linearly independent columns. Hence there is a subset S of n^2 elements of G for which the matrices $[\bar{\xi}_{ij}(x)] \in M(n, F_\nu)$ $(x \in S)$ are linearly independent over F_ν. Hence dim $F_\nu \bar{G} \geqslant n^2$, and so $\bar{G}$ is absolutely irreducible by Theorem 2.7B. ■

Exercise. Let G be an absolutely irreducible subgroup of $GL(n, F)$ where F is a field extension of finite degree over **Q**. If G is finite of order g, show that $n | g$. [*Hint:* From the proof above, $g/n \in D_\nu$ for all primes p.]

§3.8. We now summarize and combine some of the results of §3.4–3.7. Let G be a group of order g.

(R1). There is a field $F \subseteq$ **C** which is a finite degree over **Q** such that each irreducible representation of G over **C** is equivalent to one over F (Corollary 3.4A). Let p be a rational prime, ν a valuation of F at p, and D_ν the valuation ring with maximal ideal M_ν(see §3.6). Then there is a complete set of pairwise inequivalent irreducible representations $\sigma_1, \ldots, \sigma_k$ of G over **C** such that $\sigma_i : G \to GL(n_i, D_\nu)$ for each i (Theorem 3.6A). From elementary character theory we know that the number k of these representations equals the number of classes of conjugate elements in G (for example, see Theorem 16.5.5 of M. Hall's book.).

(R2). $F_\nu = D_\nu/M_\nu$ is a field of characteristic p (see (V3) of §3.6). The mapping (mod M_ν) maps G^{σ_i} into $GL(n_i, F_\nu)$, and the composite of this mapping with σ_i gives a representation $\bar{\sigma}_i : G \to GL(n_i, F_\nu)$. We shall denote the characters of σ_i and $\bar{\sigma}_i$ by χ_i and $\bar{\chi}_i$ respectively. The ordinary character relations (see the reference in (R1)) give

$$\sum_{x \in G} \chi_i(x)\, \chi_j(x^{-1}) = g\, \delta_{ij} \quad (i,\ j = 1, \ldots, k)$$

and so, writing $\bar{g}$ for g (mod M_ν) we conclude

$$\sum_{x \in G} \bar{\chi}_i(x)\, \bar{\chi}_j(x^{-1}) = \bar{g}\, \delta_{ij} \quad (i, j = 1, \ldots, k). \tag{3.8.1}$$

(R3). If $p \nmid g$, then $\bar{g} \neq 0$ by (V3) of §3.6. Thus by (3.8.1) $\bar{\chi}_i \neq \bar{\chi}_j$ if $i \neq j$. By Theorem 3.7 all σ_i are absolutely irreducible. Thus we have k pairwise inequivalent absolutely irreducible representations $\bar{\sigma}_1, \ldots, \bar{\sigma}_k$ for G over F_ν; Corollary 2.7A shows that this is a complete set.

We can now prove the following important theorem.

THEOREM 3.8. *Let E be an algebraically closed field of characteristic $p > 0$ and let G be a group of order g with $p \nmid g$. Let F be an extension field of finite degree over* **Q** *in* **C** *such that each irreducible representation of G over* **C** *is equivalent to one over F (see Corollary 3.4A). Let ν be a valuation of F at p, and D_ν the valuation ring with maximal ideal M_ν. Then, for each representation*

$\bar{\rho} : G \to GL(n, E)$ *of* G *over* E, *there exists a representation* $\sigma : G \to GL(n, D_\nu)$ *such that the composite of* σ *with the mapping* (mod M_ν) *gives a representation* $\bar{\sigma} : G \to GL(n, D_\nu/M_\nu)$ *equivalent to* $\bar{\rho}$ *(identifying* D_ν/M_ν *with its isomorphic copy in* E*). Moreover, if* $\bar{\rho}$ *is absolutely irreducible, then* σ *is absolutely irreducible.*

Proof. By Theorem 2.3 $\bar{\rho}$ is completely reducible, and its irreducible components are absolutely irreducible because E is algebraically closed (Theorem 2.7B). Thus it is enough to consider the case where $\bar{\rho}$ is absolutely irreducible. In this case $\bar{\rho}$ is equivalent to one of the representations $\bar{\sigma}_i$ in (R3) above (when we identify D_ν/M_ν with its copy in E). Then $\sigma = \sigma_i$ has the properties required. ■

We restate part of the theorem as a corollary.

COROLLARY 3.8. *Let* G *be a subgroup of order* g *in* $GL(n, E)$ *where* E *is a field of characteristic* $p > 0$. *If* $p \nmid g$, *then* G *is isomorphic to a subgroup of* $GL(n, \mathbf{C})$; *and if* G *is absolutely irreducible, then it is isomorphic to an irreducible subgroup of* $GL(n, \mathbf{C})$. ■

§3.9. *Notes and References.*

The ideas of §3.3 were first applied to representation theory in a paper originally published in 1940:

Mal'cev, A. I. On the faithful representation of infinite groups by matrices, *Am. Math. Soc. Transl.* (2), **45**, 1–18 (1965).

Our proof of Theorem 3.3B follows a paper by Artin and Tate.

The proofs (but not the theorems) of §3.4 seem new. Alternative proofs may be found in the book of Curtis and Reiner §§29, 41 and 69. There it is also shown that the field E of Corollary 3.4A may be taken as the field generated over its prime subfield by a primitive eth root of 1 where e is the exponent of G.

Exercise 2 of §4 was shown to me by L. G. Kovacs.

A short proof of Theorem 3.5 may be given using the result that a finitely generated module of a PID is a direct sum of cyclic modules.

The material on valuations in §3.6 is classical. This approach to Theorem 3.8 seems simpler than the original one such as given in

Speiser, A. *Die Theorie der Gruppen von endlicher Ordnung* (3rd ed.), Springer, Berlin (1937).

We have however followed Speiser in §3.7. Also see Huppert's book referred to in §0.4.

For a generalization of Theorem 3.8 see the proof of Corollary 3 of

Glauberman, G. Correspondences of characters for relatively prime operator groups, *Can. J. Math.* **20**, 1465–1468 (1968).

There seems a lot still to learn about when a representation may be 'lifted' in this manner from a field of characteristic p to one of characteristic 0. Chapter XI of Curtis and Reiner's book gives some further material.

CHAPTER 4

Primitivity

§4.1. Let G be an irreducible linear group over a vector space V of dimension n. Then G is called *imprimitive* if for some integer $m \geqslant 2$ there are subspaces $V_1, \ldots, V_m$ of V such that we have a direct sum $V = \bigoplus_{i=1}^{m} V_i$ and, for each $x \in G$ the mapping $V_i \mapsto V_i x$ is a permutation of the set $\{V_1, \ldots, V_m\}$. If G is not imprimitive, then it is called *primitive*.

Note. We only discuss primitivity when a group is irreducible.

Example. Let G be the (irreducible) subgroup of $GL(2, \mathbf{C})$ generated by

$$\begin{bmatrix} 0 & 1 \\ -1 & 0 \end{bmatrix} \text{ and } \begin{bmatrix} i & 0 \\ 0 & -i \end{bmatrix}.$$

If V is the underlying vector space, and V_1 and V_2 are the subspaces spanned by $(1\ \ 0)$ and $(0\ \ 1)$, respectively, then G permutes $\{V_1, V_2\}$ and so G is imprimitive.

The first part of this chapter studies the properties of primitive and imprimitive groups. In §4.2 we consider the classical relation between imprimitivity and the normal structure of the group. In §4.3–4.5 we prove a structure theorem due to Suprunenko for primitive groups of a certain type, and this result will be the main tool in our study of solvable linear groups. Finally §4.6 deals with monomial groups which will be important in our study of nilpotent linear groups.

§4.2 Imprimitivity of linear groups is directly related to the existence of certain normal subgroups. The two theorems of this section describe this relation.

Let N be a subgroup of $GL(V)$. Let W be a minimal N-space of the vector space V. Then the *homogeneous N-space V_W* is the subspace of V formed as the sum of all minimal N-spaces W' which are N-isomorphic to W.

Note. 1. By an argument similar to that of the proof of Theorem 2.2, $V_W = \bigoplus_{i=1}^{s} W_i$ where $W_1, \ldots, W_s$ are N-spaces of V which are N-isomorphic to W. Since W is a minimal N-space, V_W has an N-composition series

$$V_W = \bigoplus_{i=1}^{s} W_i \supset \bigoplus_{i=1}^{s-1} W_i \supset \ldots \supset W_1 \supset 0$$

with all composition factors N-isomorphic to W. By the Jordan–Hölder theorem (see §0.2), each minimal N-space in V_W is N-isomorphic to W.

2. If V has a N-composition series of length k, then by the Jordan–Hölder theorem there are at most k classes of N-isomorphic minimal N-space, and hence at most k homogeneous N-spaces.

THEOREM 4.2A. *Let G be an irreducible subgroup of $GL(V)$, and let N be a normal subgroup of G. Let $V_1, \ldots, V_m$ be the different homogeneous N-spaces of V, and let $W_1, \ldots, W_m$ be corresponding (pairwise non-isomorphic) minimal N-spaces. Then $V = \bigoplus_i V_i$ and each $x \in G$ permutes the set of V_i by $V_i \mapsto V_i x$. In particular, if $m > 1$, then G is imprimitive.*

Proof. Since N is completely reducible by Theorem 2.2, the definition of homogeneous N-space shows $V = \sum_{i=1}^{m} V_i$. We shall show by induction on s that

$$\sum_{i=1}^{s} V_i = \bigoplus_{i=1}^{s} V_i; \text{ that is } U_s \overset{\text{def}}{=} \bigoplus_{i=1}^{s-1} V_i \cap V_s = 0.$$

Indeed, if the N-space $U_s \neq 0$ it would contain a minimal N-space W. Since the N-composition factors for $\bigoplus_{i=1}^{s-1} V_i$ are all N-isomorphic to one of $W_1, \ldots, W_{s-1}$, the Jordan–Hölder theorem shows that W is N-isomorphic to some W_j $(1 \leqslant j < s)$. On the other hand, as an N-space in V_s, W is N-isomorphic to W_s. This contradicts the fact that all the V_i are different. Hence the induction step is proved and we have $V = \bigoplus_{i=1}^{m} V_i$.

Finally we show that for each $x \in G$ and each i, $V_i x \subseteq V_j$ for some j; then the fact that x has an inverse shows that $V_i \leftrightarrow V_i x$ is is a permutation of the V_i's. Now $V_i x$ is an N-space because $V_i xy = V_i xyx^{-1}. x = V_i x$ for all $y \in N$ by the normality of N. Similarly, if W and W' are both minimal N-spaces in $V_i x$ then Wx^{-1} and $W'x^{-1}$ are both minimal N-spaces in V_i and so both are N-isomorphic to W_i; this in turn shows that both W and W' are N-isomorphic to the minimal N-space $W_i x$. Because $V_i x$ is completely reducible as an N-space by Theorem 2.2, we conclude that $V_i x$ is contained in the homogeneous N-space V_j corresponding to the minimal N-space $W_i x$. ■

The following Corollary is an important special case.

COROLLARY 4.2A. *Let G be an irreducible linear group over a vector space V where V has dimension $n > 1$ over an algebraically* *closed field. If G is primitive then each normal abelian subgroup A of G is in Z(G).*

Proof. $V = \bigoplus_{i=1}^{n} W_i$ where the W_i are 1-dimensional A-spaces in V (Theorem 2.2 and Corollary 2.6). If G is primitive, then all the W_i must be A-isomorphic. But this means that A is a group of scalars, so $A \subseteq Z(G)$. ■

THEOREM 4.2B. *Let G be an (irreducible) imprimitive subgroup of GL(V), and let* $V = \bigoplus_{i=1}^{m} V_i$ *where the set* $\{V_1, \ldots, V_m\}$ *of m subspaces is permuted by G (and m > 1). Define* $G_i = \{x \in G | V_i x = V_i\}$ *for* $i = 1, \ldots, m$ *and put* $N = \bigcap_i G_i$. *Let* x^θ *denote the permutation* $V_i \mapsto V_i x$ $(i = 1, \ldots, m)$. *Then* $\theta : x \mapsto x^\theta$ *is a homomorphism of G onto a transitive permutation group of degree m with* $\ker \theta = N$. *In particular, all* G_i *are conjugate in G, the index* $|G : G_i| = m$, *and* $\dim V_i = d$ *for all i with* $d = n/m$. *Moreover,* V_i *is a minimal* G_i*-space for each i; hence by Theorem 2.2* $N|V_i$ *is completely reducible and N is isomorphic to a subgroup of the direct product* $N|V_1 \times \ldots \times N|V_m$.

Proof. It is easily checked that θ is a homomorphism with kernel N, and G^θ is transitive because G is irreducible. Now G_i^θ is the *stabilizer* of V_i in G^θ (that is, the set of elements of G^θ which leave the symbol V_i fixed). Because G^θ is transitive of degree m, standard theorems in the theory of permutation groups show that the stabilizers G_i^θ are all conjugate in G^θ, and $|G^\theta : G_i^\theta| = m$. Thus the inverse images G_i are all conjugate in G and $|G : G_i| = m$ because $\ker \theta = N \subseteq G_i$. Also, because G^θ is transitive, we can find $x_1, \ldots, x_m$ in G such that $V_i = V_1 x_i$ and so $\dim V_i = \dim V_1 = d$ and $n = \dim V = \Sigma \dim V_i = md$.

Next we show that V_i is a minimal G_i-space. Let $y_1, \ldots, y_m$ be a set of right coset representatives for G_i in G, and let W be a minimal G_i-space in V_i. Then $W' = \sum_{j=1}^{m} Wy_j$ is a G-space in V because for each $x \in G$ we have for each j, $y_j x \in G_i y_k$ for some k and then $Wy_j x \subseteq Wy_k$. By the irreducibility of G, $W' = V$ and so $n \leqslant m \dim W \leqslant m \dim V_i = n$; hence $W = V_i$. This shows that V_i is a minimal G_i-space. Finally N is isomorphic to a subgroup of $N|V_1 \times \ldots \times N|V_m$ by the mapping $x \mapsto (x|V_1, \ldots, x|V_m)$. ∎

Exercise. Let G be an imprimitive linear group of prime degree p. Show that G has a normal abelian subgroup A such that G/A is isomorphic to a transitive permutation group of degree p; in particular, $p \mid |G:A|$ and $|G:A| \mid p!$.

§4.3 In this section we describe the structure of irreducible linear groups which are nilpotent of class 2. This theorem has several applications, and will be used to prove the theorem of the next section.

THEOREM 4.3. *Let $\rho : G \to GL(V)$ be an irreducible representation of a group G over a vector space V of dimension n over an algebraically closed field F. If $G' \subseteq Z(G)$, then $|G : Z(G)| = n^2$, and n is prime to char F (if the latter is non-zero).*

Proof. If $x \in G$, $x \notin Z(G)$, then there exists $y \in G$ such that $xy \neq yx$. Since $x^{-1}y^{-1}xy \in Z(G)$ by hypothesis, $(x^\rho)^{-1}(y^\rho)^{-1}x^\rho y^\rho = \zeta 1$ for some $\zeta \in F$ by Corollary 2.6. Thus $\operatorname{tr} x^\rho = \operatorname{tr}(y^{-1}xy)^\rho = \operatorname{tr}(\zeta x^\rho) = \zeta \operatorname{tr} x^\rho$; and so $\operatorname{tr} x^\rho = 0$ because $\zeta \neq 1$. Hence $\operatorname{tr} x^\rho = 0$ for all $x \notin Z(G)$. On the other hand, if $x \in Z(G)$, then $x^\rho = \xi 1$ for some $\xi \in F$ and so $\operatorname{tr} x^\rho = n\xi$ (with $\xi \neq 0$). From Theorem 2.7A we cannot have $\operatorname{tr} x^\rho = 0$ for all $x \in G$, so $n \neq 0$ (that is, n is prime to the characteristic of F) and $\operatorname{tr} x^\rho \neq 0$ if and only if $x \in Z(G)$.

Theorem 2.7B shows that $FG^\rho = \mathrm{Hom}(V, V)$ and so we can find x_i^ρ $(i = 1, \ldots, n^2)$ in G^ρ as a basis for FG^ρ. We shall finish the proof of the theorem by proving that $x_1, \ldots, x_{n^2}$ is a set of coset representatives for $Z(G)$ in G. Indeed, since $z^\rho x_i^\rho$ is a scalar multiple of x_i^ρ for each $z \in Z(G)$, the cosets $Z(G)x_i$ are different for different i by the linear independence of the x_i^ρ. It remains to show that each $x \in G$ lies in some $Z(G)x_i$ $(i = 1, \ldots, n^2)$. Because $x^\rho \in FG^\rho$, $x^\rho = \sum \xi_i x_i^\rho$ for some $\xi_i \in F$ where ξ_j, say, is not zero. Now we know that $\mathrm{tr}\, y^\rho \neq 0$ for $y \in G$ only if $y \in Z(G)$. Therefore

$$\mathrm{tr}(xx_j^{-1})^\rho = \sum \xi_i \,\mathrm{tr}(x_i x_j^{-1})^\rho = \xi_j \,\mathrm{tr}\, 1 \neq 0$$

and hence $xx_j^{-1} \in Z(G)$. This completes the proof. ■

Exercises. 1. Let N be a finite normal subgroup of an irreducible linear group G of degree $n > 1$. Show that $\sum_{x \in N} x = 0$.

[*Hint:* First reduce to the case of an algebraically closed field, and then use Corollary 2.6.]

2. With the notation of Exercise 1, show that if $y \in G$ and the centralizer $C_N(y)$ of y in N is 1, then $|N|\ \mathrm{tr}\, y = 0$.
[*Hint:* $N = \{y^{-1}x^{-1}yx \mid x \in N\}$.]

3. Show that if the hypothesis that F is algebraically closed is replaced by 'F is perfect' in Theorem 4.3 we can conclude '$|G : Z(G)| = d^2$ for some divisor d of n'. [*Hint:* Use Theorems 2.10A and 2.10B.]

§4.4 The theorem of this section is a powerful structure theorem for a certain class of primitive groups. It will be the main tool used in analysing the structure of solvable linear groups (Chapter 6). It is essentially due to D. A. Suprunenko (1956).

The Fitting subgroup Fit(G) of a group G is the subgroup generated by the set of all normal nilpotent subgroups of G. Clearly Fit(G) is a characteristic subgroup of G.

Note. 1. A theorem of *H*. Fitting (see p. 257 of Huppert's book referred to in §0.4) shows that if M and N are normal nilpotent subgroups of G, then so is MN. Thus each finite subset of Fit(G) is contained in a normal nilpotent subgroup of G. In particular, Fit(G) is locally nilpotent.

2. If G is a non-abelian solvable group, $G/Z(G)$ has a non-trivial normal abelian subgroup $N/Z(G)$. Since $N \subseteq$ Fit(G), Fit(G) $\supset Z(G)$.

THEOREM 4.4. *Let V be a vector space of dimension n over an algebraically closed field F. Let G be a primitive subgroup of $GL(V)$ such that* Fit(G) $\supset Z(G)$. *Then:*

(1) *$Z(G)$ is the unique maximal normal abelian subgroup of G. ($Z(G)$ is a group of scalars by Corollary 2.6.)*

(2) Fit(G)/$Z(G)$ *is the unique maximal normal abelian subgroup of $G/Z(G)$.*

(3) $|\text{Fit}(G): Z(G)| = d^2$ *for some divisor d of n; $d = n$ if* Fit(G) *is irreducible.*

(4) *The Sylow subgroups of* Fit(G)/$Z(G)$ *are elementary abelian.*

(5) *Let $d = p_1^{\ell_1} \ldots p_s^{\ell_s}$ be the canonical prime decomposition of d* (see (3)). *Then there is a homomorphism θ of G into the direct product of the symplectic groups $Sp(2\ell_i, p_i)$ ($i = 1, \ldots, s$) with* $\ker\theta = \{x \in G \mid [x, u] \in Z(G)$ *for all* $u \in \text{Fit}(G)\} = C_G(\text{Fit}(G)/Z(G))$.

Proof. Write $Z = Z(G)$ and $H =$ Fit(G).

(1) This follows from Corollary 4.2A.

(2) For each normal abelian subgroup A/Z of G/Z we have

$A \subseteq H$. Therefore it is enough to prove H/Z is abelian. From Note 1 above it is enough to prove that for each normal nilpotent subgroup $N \supseteq Z$ of G, N/Z is abelian. Let $N = \gamma_1(N) \supset \gamma_2(N) \supset \ldots \supset \gamma_{k+1}(N) = 1$ be the lower central series for N. Then $[\gamma_i(N), \gamma_j(N)] \subset \gamma_{i+j}(N)$ for all i, j and so $\gamma_i(N)$ is abelian if $i > (½)(k+1)$. Then $\gamma_i(N)$ and $Z(N)$ are normal abelian subgroups of G and so $\gamma_i(N) \subseteq Z = Z(N)$ by (1); hence $i > k$. Thus $i > ½(k+1)$ implies $i > k$, and hence $k \leqslant ½(k+1) + ½$, so $k \leqslant 2$. Thus $N' \subseteq Z(N) = Z$ and N/Z is abelian as required.

(3) Let W be a minimal H-space in V and put $d = \dim W$; $d|n$ by Theorem 2.2. Then $x \mapsto x|W$ is an irreducible representation of degree d for H. Since $Z(H) = Z$ by (1), we have from (2) and Theorem 4.3 that $|H : Z| = |H : Z(H)| = d^2$.

(4) Let P/Z be a non-trivial Sylow p-group of H/Z and suppose it has exponent p^e; we must show $e = 1$. For each integer $f \geqslant e/2$ and all $x, y \in P$ we have $[x, y] \in Z$ and so $[x^{p^f}, y^{p^f}] = [x, y^{p^f}]^{p^f} = [x, y^{p^{2f}}] = 1$ because $y^{p^{2f}} \in Z$. Since H/Z is abelian, P is a characteristic subgroup of H and so P is normal in G. Thus $P_0 = \{x^{p^f} \mid x \in P\}$ generates a normal abelian subgroup of G. From (1) we conclude $x^{p^f} \in Z$ for all $x \in P$, so $f \geqslant e$. Thus each integer $f \geqslant e/2$ is $\geqslant e$. Hence $e/2 + 1/2 \geqslant e$ and $e \leqslant 1$ as required.

(5) Let P/Z be a non-trivial Sylow p-group of H/Z; then P/Z is an elementary abelian group of order $p^{2\ell}$ where p^{ℓ} is the power of p dividing d (see (3) and (4)). Thus there is a group isomorphism $Zx \mapsto \bar{x}$ of P/Z onto the additive group of a vector space $\bar{P}$ of dimension 2ℓ over the field $GF(p)$ of p elements. We now define an alternating form f on $\bar{P}$ (see §1.5). Let ζ be a fixed primitive pth root of 1 in F. For all $x, y \in P$, $[x, y] \in Z$ and so $[x, y]^p = [x, y^p] =$ by (4); hence $[x, y] = \zeta^i.1$ for some integer i, $0 \leqslant i \leqslant p-1$. Identifying i with the corresponding element of $GF(p)$, we define

$f : \bar{P} \times \bar{P} \to GF(p)$ by $f(\bar{x}, \bar{y}) = i$ when $[x, y] = \zeta^i \cdot 1$. It is readily verified that f is a non-degenerate alternating form on $\bar{P}$. For example, $f(\bar{x}, \bar{y} + \bar{z}) = f(\bar{x}, \bar{y}) + f(\bar{x}, \bar{z})$ because $[x, yz] = [x, z]z^{-1}[x, y]z = [x, y][x, z]$ for all $x, y, z \in P$, and f is non-degenerate because for each $x \in P$, $x \notin Z$ there is a $y \in P$ such that $[x, y] \neq 1$.

Now consider the action of G on P/Z by conjugation; note that P is normal in G since it is characteristic in H. For each $u \in G$ we define $u_P \in GL(\bar{P})$ by $u_P : \bar{x} \mapsto \overline{u^{-1}xu}$. Then u_P leaves f invariant because $[u^{-1}xu, u^{-1}yu] = u^{-1}[x, y]u = [x, y]$ for all $x, y \in P$. Thus $u \mapsto u_P$ is a homomorphism of G into $Sp_f(\bar{P}) \cong Sp(2\ell, p)$.

Finally the mapping $\theta : u \mapsto (u_{P_1}, \ldots, u_{P_s})$ gives a homomorphism of G into $\bigotimes_{i=1}^{s} Sp(2\ell_i, p_i)$ with kernel $C_G(H/Z)$ as required. ■

Exercises. 1. Show that the conclusions of Theorem 4.4 remain valid when the hypothesis 'G is primitive' is weakened to 'G is irreducible and $Z(G)$ is the unique maximal normal abelian subgroup of G'.

2. Show that the action of G on $\bar{P}$ described in (5) has no 1-dimensional invariant subspaces. In particular, if $\ell = 1$, then the induced group is an irreducible subgroup of $Sp(2, p)$.

3. If $s = 1$ and $p_1 = p$ in (5), show that $G/\mathrm{Fit}(G)$ has no non-trivial normal p-subgroup.

4. If G is finite with an abelian Sylow p-group for some prime p, show that $p \nmid d$ where d is defined in (3). [*Hint:* Use (1).]

§4.5 Theorem 4.4 is particularly important for solvable groups.

THEOREM 4.5. *Let G be a solvable primitive subgroup of $GL(V)$ where V has dimension $n > 1$ over an algebraically closed* *field. Then* Fit(G) $\supset Z(G)$ *and so conclusions* (1)–(5) *of Theorem* 4. *hold. In addition*

$$\text{(6)} \qquad \text{Fit}(G) = C_G(\text{Fit}(G)/Z(G))$$

and so (5) *describes the structure of* $G/\text{Fit}(G)$.

Remark. (3), (5) and (6) show that $|G : Z(G)| < \infty$.

Proof. $G/Z(G)$ is a non-trivial solvable group because $n > 1$, so it has a normal abelian subgroup $A/Z(G) \neq 1$. Then $\text{Fit}(G) \supseteq A \supset Z(G)$.

Now suppose (6) is false. Then $C = C_G\,(\text{Fit}(G)/Z(G))$ is a normal subgroup of G properly containing Fit(G) by (2). Since G is solvable, there is a normal abelian subgroup $A/\text{Fit}(G) \neq 1$ of $G/\text{Fit}(G)$ contained in $C/\text{Fit}(G)$. Then $A' \subsetneq \text{Fit}(G)$ and $[A, A'] \subseteq Z(G)$ because $A \subseteq C$. This implies that A is nilpotent and so $A \subseteq F(G)$ contrary to the choice of A. Thus $C = \text{Fit}(G)$ as asserted

Exercises. 1. Let G be a solvable primitive linear group of prime degree p over an algebraically closed field. Prove that $|G : \text{Fit}(G)|$ divides $p(p^2 - 1)$. [*Hint:* Use Theorem 1.5.]

2. Let F be an algebraically closed field. Show that each solvable primitive subgroup of $SL(n, F)$ is finite. [*Hint:* Show that the center has order $\leqslant n$.]

§4.6 The extreme form of imprimitivity occurs for a monomial group.

DEFINITION. A subgroup G of $GL(V)$ where V is a vector space of dimension n over a field F is *monomial* if there exist n 1-dimensional subspaces $V_1, \ldots, V_n$ of V such that $V = \bigoplus_{i=1}^{n} V_i$ and

$V_i \mapsto V_i x$ is a permutation of the set $\{V_1, \ldots, V_n\}$ for each $x \in G$. (We shall *not* insist that G is irreducible.)

Note. For each V_i choose a non-zero $v_i \in V_i$. Then $v_1, \ldots, v_n$ is a basis for V and over this basis G corresponds to a subgroup of Mon(n, F), the monomial group of matrices (see §1.3). In particular $N = \{x \in G | V_i x = V_i \text{ for } i = 1, \ldots, n\}$ is a normal abelian subgroup of G and G/N is isomorphic to a permutation group of degree n; when G is irreducible this permutation group is transitive (compare Theorem 4.2A).

Since monomial groups have a relatively simple structure, the following theorems are of interest. We shall call a representation *primitive* or *monomial* if its image is primitive or monomial, respectively.

THEOREM 4.6A. *Let* Γ *be a class of groups with the property that if G is in* Γ *then each subgroup and each factor group of G is in* Γ. *Let F be a given field. Then:*

Each completely reducible representation over F of each G in Γ *is a monomial representation* $\Longleftrightarrow$ *Each primitive representation over F of each G in* Γ *is of degree* 1.

Proof. ($\Longrightarrow$) Trivial because a primitive group of degree > 1 is not monomial.

($\Longleftarrow$) Let G be in Γ and let $\rho : G \to GL(V)$ be a completely reducible representation of degree n over F. We must show that ρ is a monomial representation. This is true if $n = 1$, so suppose $n > 1$ and proceed by induction on n. If G^ρ is reducible, then the result follows by complete reducibility and induction; hence suppose G is irreducible. By hypothesis G^ρ is imprimitive, so we can write $V = \bigoplus_{i=1}^{m} V_i$ ($m \geqslant 2$) where the subspaces V_i are permuted by G^ρ. Theorem 4.2B shows that G^ρ has a subgroup of G_1^ρ of index m for which V_1 is a minimal G_1^ρ-space. By induction and the hypothesis on Γ we conclude that G_1^ρ is a monomial group, and so $V_1 = \bigoplus_{j=1}^{d} W_j$ where the W_j are 1-dimensional subspaces of V_1 which are permuted by the elements of G_1^ρ and $d = n/m$.

Now let $x_1^\rho, \ldots, x_i^\rho$ be right coset representatives for G_1^ρ in G^ρ. Then the set of n 1-dimensional subspaces $W_j x_i^\rho$ ($j = 1, \ldots, d$; $i = 1, \ldots, m$) is permuted by elements of G^ρ. Indeed, for $x^\rho \in G^\rho$ and each i there is some $y^\rho \in G_1^\rho$ and some k such that $x_i^\rho x^\rho = y^\rho x_k^\rho$, and so $(W_j x_i^\rho)x^\rho = (W_j y^\rho)x_k^\rho = W_\ell x_k^\rho$ for some ℓ. In particular, this shows that $\sum_{i,j} W_j x_i^\rho$ is a G^ρ-space in V, and so is equal to V by the irreducibility of ρ. Finally, counting dimensions we conclude $V = \bigoplus_{i,j} W_j x_i^\rho$ and the proof of the theorem is complete. ■

We now give an application of Theorem 4.6A.

THEOREM 4.6B. *Let Γ be a class of groups such that if G is in Γ then each subgroup and each factor group of G is in Γ. Suppose that each non-abelian group G in Γ has a normal abelian subgroup which is not contained in Z(G). Then each completely reducible representation over an algebraically closed field of each G in Γ is a monomial representation.*

In particular, the following classes satisfy our hypothesis:

(1) *The class of all nilpotent groups.*

(2) *The class of all metabelian groups (solvable of length $\leqslant 2$).*

(3) *The class of all finite solvable groups G with a normal subgroup N such that G/N is supersolvable and N has all its Sylow subgroups abelian.*

Proof. The first assertion follows from Theorem 4.6A and Corollary 4.2A. Since each of the classes (1)–(3) are clearly closed under taking subgroups and factor groups, it remains to show that each of these classes satisfies the condition on the existence of non-central normal abelian subgroups.

(1) If G is a non-abelian nilpotent group, and $Z = Z(G)$, then we can take $x \in G$, $x \notin Z$ such that xZ is in the center of G/Z. Then $A = \langle x, Z \rangle$ is a normal abelian subgroup of G and $A \supset Z$.

(2) If G is a non-abelian metabelian group, then $G' \neq 1$ and $G'' = 1$. If $Z(G) \supseteq G'$, then G is nilpotent and we have already considered this case. If $Z(G)$ does not contain G', then G' is the required normal abelian subgroup.

(3) Let G be a non-abelian group of the type described, and put $H = \mathrm{Fit}(N)$. Since N has all its Sylow subgroups abelian, and H is nilpotent, therefore H is abelian. As a characteristic subgroup of N, H is normal in G. If H is not contained in $Z(G)$ it is the subgroup we are looking for; so suppose $H \subseteq Z(G)$. Then because H is the largest nilpotent subgroup of N, $H = Z(G) \cap N$. If $H \neq N$, then the solvable group G/H has a normal abelian subgroup $K/H \neq 1$ contained in the normal subgroup N/H. But then $K' \subseteq H \subseteq Z(G)$ so K is a normal nilpotent subgroup of N properly containing H. This contradicts the definition of H, so $H = N$ and $N \subseteq Z(G)$. Then $G/Z(G)$ is a quotient group of G/N and hence supersolvable. Since the minimal normal subgroups of a finite supersolvable group are cyclic of prime order, G has a normal subgroup A with $A/Z(G) = \langle xZ(G) \rangle$ cyclic and so $A = \langle x, Z(G) \rangle$. In this case A is is the required normal abelian subgroup of G. ■

A group is *locally nilpotent* if each finite subset generates a nilpotent subgroup. We shall later show that every locally nilpotent linear group is solvable (Corollary 6.2B). Using that result we have the following corollary.

COROLLARY 4.6B. *If G is a locally nilpotent group, then each completely reducible representation of G over an algebraically closed field is a monomial representation.*

Proof. The class of all locally nilpotent groups satisfies the hypotheses of Theorem 4.6A. Therefore it follows from that theorem that it is enough to show that each primitive representation ρ of G over an algebraically closed field is of degree 1. By Corollary 6.2B, G^{ρ} is solvable, and so $|G^{\rho}: Z(G^{\rho})| < \infty$ by Theorem 4.5. Since $G^{\rho}/Z(G^{\rho})$ is a finite locally nilpotent group it is nilpotent, and therefore G^{ρ} is nilpotent. Then Theorem 4.6B (1) shows that G^{ρ} is a monomial group. But G^{ρ} is primitive, so ρ has degree 1. ■

Exercises. 1. Let G be a finite solvable irreducible linear group of degree n over an algebraically closed field. If, for each prime $p|n$, the Sylow p-groups of G are abelian, prove that G is monomial. [*Hint:* Use Theorem 4.5.]

2. Let G be a completely reducible linear group of degree n. If G is a finite p-group, show that G has a normal abelian subgroup of index p^m where

$$m \leqslant [n/p] + [n/p^2] + \ldots < n/(p-1).$$

§4.7. *Notes and References*

The results of §4.2 are classical. Theorem 4.3 is more recent and has been proved independently by several authors. The Exercises of §4.3 generalize slightly an important lemma of Feit and Thompson.

The central theorem of this chapter is Theorem 4.5. It appears as Theorem 11 in

Suprunenko, D. A. Soluble and Nilpotent Linear Groups, *Transl. Math. Monographs*, Vol.9, Amer. Math. Soc., Rhode Island (1963)

The case of a general field is dealt with there. Our proof is quite different from Suprunenko's proof. Similar results are found in

Rigby, J. F. Primitive linear groups, *J. Lond. math. Soc.* **35**, 389–400 (1960).

Theorem 4.6B for the case of a finite *p*-group was proved by Blichfeldt in 1904 (he also used Corollary 4.2B). Case (3) of that theorem is due to Huppert, and Corollary 4.6B is from the book of Suprunenko referred to above. The statement of Theorem 4.6A seems to be new but the proof is essentially due to Zassenhaus. Quite a lot is known about which finite groups have all their representations over **C** as monomial representations; in particular such groups are solvable. For further details, see p. 578 of Huppert's book.

CHAPTER 5

Finite Non-Modular Groups

§5.1 A finite linear group G over a field F is *non-modular* if either char $F = 0$ or char $F = p > 0$ but $p \nmid |G|$; otherwise G is *modular*. This chapter deals with the structure of finite non-modular groups. It turns out that we really only have to deal with linear groups over **C**. This follows from the next theorem which in turn follows immediately from Theorem 3.4B and Corollary 3.8.

THEOREM 5.1. *If G is a finite non-modular linear group of degree n, then G is isomorphic to a subgroup G^* of $GL(n, \mathbf{C})$. Moreover, if G is absolutely irreducible, we may take G^* to be (absolutely) irreducible.* ■

In §5.2 we have results on the degrees of absolutely irreducible groups. There are then some arithmetical results on the value of characters of finite groups and these are applied in §5.4 and §5.5 to prove criteria for the existence of certain Hall subgroups and of normal Sylow p-groups for large p. In §5.6 we have the classical theorem that each representation over **C** of a finite group is equivalent to a unitary representation. The chapter concludes with Jordan's theorem: For each integer $n > 0$ there is a bound $\beta(n)$ such that each finite non-modular linear group of degree n has a normal abelian subgroup of index $\leqslant \beta(n)$. Perhaps we should point out that §§5.2, 5.6 and 5.7 are independent of the rest and it is only these results that we use in later chapters.

§5.2 The theorem of this section is due to N. Itô (1951). It generalizes the following classical result of Schur which uses some elementary character theory.

LEMMA 5.2. *Let G be a finite irreducible subgroup of* $GL(n, \mathbf{C})$. *Then n divides* $|G:Z(G)|$.

Proof. Let χ be the character of $G : \chi(x) = \operatorname{tr} x \ (x \in G)$. Let $C_1 = \{1\}, C_2, \ldots, C_k$ be the classes of conjugate elements in G and put $h_i = |C_i|$. Since $\chi(x)$ only depends on the class C_i to which x belongs we shall write $\chi(x) = \chi_i$ when $x \in C_i$. If $z \in Z(G)$, then zC_i is a conjugacy class and so we can define an equivalence relation on the set of conjugacy class by: $C_i \sim C_j \Longleftrightarrow C_i = z\, C_j$ for some $z \in Z(G)$. Since G is irreducible, each $z \in Z(G)$ is a scalar by Corollary 2.6. Thus, if $x, y \in C_i$ and $x = zy$ for some $z \in Z(G)$, then $\chi_i = \chi(x) = \chi(zy) = \zeta\, \chi(y) = \zeta\, \chi_i$ for some $\zeta \in \mathbf{C}$, $\zeta \neq 0$; hence either $z = 1$ or $\chi_i = 0$. This shows that if $\chi_i \neq 0$, then the equivalence class containing C_i contains $|Z(G)|$ different conjugacy classes. Therefore, from the elementary character relations (see, for example, Theorem 16.6.9 of M. Hall's book), and noticing that $h_i = h_j$ when $C_i \sim C_j$, we get

$$|G| = \sum_{i=1}^{k} h_i\, |\chi_i|^2 = \Sigma'\, |Z(G)|\, h_i\, |\chi_i|^2$$

where the latter sum is over a set of representatives C_i of our equivalence classes. Hence $|G : Z(G)| = \Sigma'\, h_i\, |\chi_i|^2$. But each $h_i\chi_i/n$ is an algebraic integer (see Theorem 16.8.3 of M. Hall), and so $|G : Z(G)|/n$ is an algebraic integer. Since the latter number is rational, it is a rational integer and so n divides $|G : Z(G)|$. ■

THEOREM 5.2. *Let G be a finite absolutely irreducible non-modular linear group of degree n. If A is a normal abelian subgroup of G, then n divides $|G : A|$.*

Proof. From Theorem 5.1 it is enough to consider the case where the underlying field is **C**. If $A \subseteq Z(G)$, then the theorem follows from Lemma 5.2. If A is not contained in $Z(G)$, then Corollary 4.2A shows that G is imprimitive. Then with the notation of Theorem 4.2B we conclude that G has a subgroup G_1 and a normal subgroup N such that $A \subseteq N \subseteq G_1$, $|G : G_1| = m$ and $G_1|V_1$ is an irreducible group of degree n/m $(m > 1)$. Hence by induction on the degree we conclude that n/m divides $|G_1|V_1 : A|V_1|$. But the latter divides $|G_1 : A|$, and so $n = m \cdot n/m$ divides $|G : A| = |G : G_1|\,|G_1 : A|$. ■

COROLLARY 5.2. *Let G be a finite non-modular linear group of degree n. If G has order g and each prime p dividing g is greater than n, then G is abelian.*

Proof. We may suppose that the underlying field is algebraically closed, and note that G is completely reducible by Theorem 2.3. If d is the degree of an irreducible component of G, then $d|g$ by the Theorem (taking $A = 1$). Since $p > n \geqslant d$ for all primes $p|g$, we conclude $d = 1$. Thus G corresponds to a group of diagonal matrices over a suitable basis, and so G is abelian. ■

Exercises. 1. Show that the conclusion of Theorem 5.2 remains valid when A is only assumed to be a subnormal abelian subgroup of G. ('Subnormal' is defined in §2.2).

2. Let G be a finite absolutely irreducible non-modular linear group of degree n. If G is nilpotent show that $n^2 \,\big|\, |G : Z(G)|$. [*Hint:* The Sylow p-groups of G are normal so Theorem 2.2 lets us

reduce to the case where G is a p-group and n is a power of p. Now apply the Exercise of §2.7].

§5.3 The present section gives several results on the algebraic nature of certain characters. We shall apply these results in the next section.

Let G be a finite group of order g and consider representations of G over $\mathbf{C}$. By Corollary 3.4A there is a subfield E of $\mathbf{C}$ which is a finite extension of $\mathbf{Q}$ and such that each representation of G over $\mathbf{C}$ is equivalent to one over E. By extending E if necessary, we may suppose E is a finite normal extension of $\mathbf{Q}$, and that $E \supseteq F = \mathbf{Q}(\zeta)$ where ζ is a primitive gth root of unity. From elementary Galois theory we know that each $\alpha \in \mathrm{Gal}(F/\mathbf{Q})$ (the Galois group of F over $\mathbf{Q}$) can be extended to an automorphism in $\mathrm{Gal}(E/\mathbf{Q})$; we shall choose one such automorphism of E and denote it by the same symbol α.

Let $\rho : G \to GL(n, E)$ be a representation of G affording a character χ. The values of χ lie in F because, for each $x \in G$, the eigenvalues of $\rho(x)$ are gth roots of 1. For each $\alpha \in \mathrm{Gal}(F/\mathbf{Q})$ we have a mapping $\rho^{\alpha} : G \to GL(n, E)$ where $x^{\rho} \mapsto x^{\rho^{\alpha}}$ is the mapping on G^{ρ} induced by α (see §3.1). Clearly ρ^{α} is a representation of G affording the character χ^{α} defined by $\chi^{\alpha}(x) = \{\chi(x)\}^{\alpha}$, and χ^{α} is an irreducible character if and only if χ is. Thus each $\alpha \in \mathrm{Gal}(F/\mathbf{Q})$ induces a permutation of the set of irreducible complex characters of G. It should be recalled that $\mathrm{Gal}(F/\mathbf{Q}) = \{\alpha_k \mid 1 \leqslant k \leqslant g,\ (k, g) = 1\}$ where α_k is characterized by $\zeta^{\alpha_k} = \zeta^k$ where ζ is our fixed primitive gth root of 1 (see van der Waerden's book). In particular, $|\mathrm{Gal}(F/\mathbf{Q})| = \varphi(n)$, the Euler φ-function, and $\mathrm{Gal}(F/\mathbf{Q})$ is abelian.

LEMMA 5.3A. *Let $\mathbf{Q}_h$ be the subfield of $\mathbf{C}$ generated by the primitive hth roots of 1. If $r, s > 0$ are integers with $(r, s) = d$, then $\mathbf{Q}_r \cap \mathbf{Q}_s = \mathbf{Q}_d$.*

Proof. Let m be the least common multiple of r and s. Since we are dealing with normal extensions of $\mathbf{Q}$, Galois theory shows that

$$\mathrm{Gal}(\mathbf{Q}_m/\mathbf{Q}_r \cap \mathbf{Q}_s) \simeq \mathrm{Gal}(\mathbf{Q}_m/\mathbf{Q}_r) \times \mathrm{Gal}(\mathbf{Q}_m/\mathbf{Q}_s)$$

and so

$$[\mathbf{Q}_m : \mathbf{Q}_r \cap \mathbf{Q}_s] = [\mathbf{Q}_m : \mathbf{Q}_r][\mathbf{Q}_m : \mathbf{Q}_s].$$

Therefore from the observation preceding the lemma we have

$$\varphi(m)/[\mathbf{Q}_r \cap \mathbf{Q}_s : \mathbf{Q}] = \varphi(m)^2/\varphi(r)\,\varphi(s).$$

But by the definition of m and d, we have $\varphi(m)\varphi(d) = \varphi(r)\varphi(s)$. Therefore $[\mathbf{Q}_r \cap \mathbf{Q}_s : \mathbf{Q}] = \varphi(d) = [\mathbf{Q}_d : \mathbf{Q}]$. Clearly $\mathbf{Q}_d \subseteq \mathbf{Q}_r \cap \mathbf{Q}_s$, so we conclude $\mathbf{Q}_d = \mathbf{Q}_r \cap \mathbf{Q}_s$. ∎

We shall say that an algebraic number ξ *requires* rth *roots of* 1 if $\xi \in \mathbf{Q}_r$ and $\xi \notin \mathbf{Q}_d$ for any $d < r$. Lemma 5.3A shows that when r exists it is uniquely defined.

The next lemma is due to H. Blichfeldt (1904) and we give a proof due to R. Brauer (1964).

LEMMA 5.3B. *Let G be a finite group of order g, and let χ be a complex-valued character for G. Suppose $p_1, \ldots, p_s$ are distinct primes and that there exist $x_i \in G$ such that $\chi(x_i)$ requires $p_i^{k_i}$-th roots of 1 for some $k_i > 0$ $(i = 1, \ldots, s)$. Then G has an element of order $p_1^{k_1} \ldots p_s^{k_s}$.*

Proof. Let ρ be a representation of G affording the character χ. If $\chi(x)$ requires rth roots of 1, then at least one eigenvalue of x^ρ is an hth root of 1 with $r|h$. Thus, if $\chi(x)$ requires rth roots of 1, then r divides the order of x.

Let $p_i^{m_i}$ be the largest power of p_i dividing g, and note $k_i \leqslant m_i$.

Let $q_i = g/p_i^{m_i - k_i + 1}$, and note $\chi(x_i) \notin \mathbf{Q}_{q_i}$ (with the notation of Lemma 5.3A). Thus, there exists $\alpha_i \in \mathrm{Gal}(\mathbf{Q}_g/\mathbf{Q}_{q_i})$ such that $\chi(x_i)^{\alpha_i} \neq \chi(x_i)$; but $\chi(x_j)^{\alpha_i} = \chi(x_j)$ for $j \neq i$ because $\chi(x_j) \in \mathbf{Q}_{q_i}$. If we can find $z \in G$ such that for all i, $\chi(z)^{\alpha_i} \neq \chi(z)$, then $\chi(z)$ must require mth roots of 1 where $p_i^{k_i} | m$ for all i. By the observation at the beginning of this proof, this means m divides the order of z, and hence a suitable power of z has order $p_1^{k_1} \ldots p_s^{k_s}$.

It remains to show that there exists $z \in G$ such that for all i, $\chi(z)^{\alpha_i} \neq \chi(z)$. If this were false, then (because $\mathrm{Gal}(\mathbf{Q}_g/\mathbf{Q})$ is abelian)

$$\chi(x) - \Sigma\, \chi^{\alpha_i}(x) + \Sigma\, \chi^{\alpha_i \alpha_j}(x) - \ldots = 0$$

for all $x \in G$ where the mth sum is over all $\binom{s}{m}$ sets of m distinct indices. The irreducible characters of G are linearly independent by Theorem 2.7A, so for some product $\beta = \alpha_{i_1} \alpha_{i_2} \ldots \alpha_{i_m}$ $(i_1 < i_2 < \ldots < i_m)$ we have $\chi = \chi^{\beta}$ (note m is odd). But we have seen that $\chi(x_{i_1})^{\beta} = \chi(x_{i_1})^{\alpha_{i_1}} \neq \chi(x_{i_1})$, so we have contradiction. This proves the existence of z and the lemma is proved. ∎

For our proof of the next lemma we need the fact that the cyclotomic polynomials $\Phi_n(X)$ are irreducible over $\mathbf{Q}$. A general proof is given in van der Waerden's *Modern Algebra*, Vol. 1, §53; in our case where n is prime, there is also an easy proof using the Eisenstein criterion for irreducibility.

LEMMA 5.3C. *Let G be a finite subgroup of $GL(n, \mathbf{C})$, and let χ be its character. If $x \in G$ has order p where p is prime and $p > n+1$, then $\chi(x)$ requires pth roots of* 1.

Proof. Let ζ be a primitive pth root of 1, and let d_i be the multiplicity of the eigenvalue ζ^i for x $(i = 0, 1, \ldots, p-1)$. Then

$\chi(x) = \Sigma\, d_i \zeta^i$. Now ζ is a zero of the cyclotomic polynomial $\Phi_p(X)$ over $\mathbf{Q}$. Therefore, if $\chi(x) \in \mathbf{Q}$, it would follow that $\Phi_p(X)$ divides $\Sigma\, d_i X^i - \chi(x)$ because the former is irreducible over $\mathbf{Q}$. But then $d_0 - \chi(x) = d_1 = \ldots = d_{p-1}$. Since all $d_i \geqslant 0$ and $d_0 + d_1 + \ldots + d_{p-1} = n < p-1$, we conclude $d_1 = \ldots = d_{p-1} = 0$ and so all eigenvalues of x are 1; hence $x = 1$. This contradicts the hypothesis on x. Therefore $\chi(x) \notin \mathbf{Q}$, and so $\chi(x)$ requires pth roots of 1. ■

Exercises. 1. [Generalization of Lemma 5.3C] Let G be a finite subgroup of $GL(n, \mathbf{C})$ with character χ. Let $x \in G$ have order $h \neq 1$.

(a) If $\chi(x) \in \mathbf{Q}$, show that some prime $p|h$ with $p \leqslant n+1$.

(b) If $\chi(x) = 0$, show that some prime $p|h$ with $p \leqslant n$.

[*Hint:* Write $\chi(x) = \Sigma d_i\omega_i$ where $\omega_1, \ldots, \omega_r$ are the distinct eigenvalues of x and the d_i are positive integers. If each prime $p|h$ satisfies $p > n+1$ then there exists $\alpha_k \in \mathrm{Gal}(\mathbf{Q}_h/\mathbf{Q})$ such that $\omega_i^{\alpha_k} = \omega_i^k$ $(i = 1, \ldots, r;\ k = 1, \ldots, n+1)$. If $\chi(x) \in \mathbf{Q}$, then this gives $n+1$ equations

$$\chi(x) = d_1\omega_1^k + \ldots + d_r\omega_r^k \quad (k = 1, \ldots, n+1)$$

for the $r(\leqslant n)$ integers $d_1, \ldots, d_r$.]

2. Let χ be a complex-valued character of a finite group G. Let A and B be subgroups of G and put $C = \langle A, B \rangle$ and $D = A \cap B$. For each subgroup H of G let $\chi|H$ be the restriction of χ to H and let 1_H be the unit character on H. We use $(\ ,\ )_H$ to denote the usual inner product over H. Show that

$$(\chi|C, 1_C) + (\chi|D, 1_D) \geqslant (\chi|A, 1_A) + (\chi|B, 1_B).$$

[*Hint:* If $\rho : G \to GL(V)$ is a representation affording χ, show that $(\chi|H, 1_H)$ is the dimension of the subspace

$\{v \in V \mid vx^{\rho} = v \quad \text{for all } x \in H\}$.]

3. Let $x \in GL(n, \mathbf{C})$ have order p^m (p prime) and suppose $\operatorname{tr} x^k \in \mathbf{Q}$ for $k = 0, 1, \ldots$. Let ζ be a primitive p^mth root of 1 in $\mathbf{C}$ and suppose ζ^i is an eigenvalue of x with multiplicity m_i for $i = 0, 1, \ldots, p^m-1$. Prove that $\operatorname{tr} x = n - pt$ where $t = \Sigma\, m_i \leqslant n/(p-1)$. In particular $\operatorname{tr} x \equiv n \pmod p$.

4. Let G be a finite subgroup of $GL(n, \mathbf{C})$. If the character χ of G is rational valued show that a Sylow p-group P of G must have order $\leqslant p^e$ where

$$e = \sum_{i=0}^{\infty} \left[\frac{n}{p^i(p-1)}\right]$$

(here $[s]$ is the largest integer $\leqslant s$). [*Hint*: Let P have h_t elements where the character has value $n - pt$ $(0 \leqslant t \leqslant n/(p-1))$; see Exercise 3. Show that $\Sigma\, h_t(n - pt)^i \equiv 0 \pmod{p^e}$ for $0 \leqslant i \leqslant n/(p-1)$, and hence prove $p^e \mid p^d d!$ where $d = [n/(p-1)]$.]

§5.4 The theorem of this section is due to H. F. Blichfeldt (1904) and it gives a criterion for the existence of certain Hall subgroups in linear groups. We shall need the following general lemma. Recall that a *Hall subgroup* of a finite group is a subgroup whose order and index are relatively prime.

LEMMA 5.4. *Let A and B be nilpotent Hall subgroups of the same order, say $m > 1$, in a finite group G. Then A is conjugate to B in G.*

Proof. We use induction on $|G|$. If p is a prime dividing m, and P_A and P_B are Sylow p-groups of A and B, respectively, then P_A and P_B are Sylow p-groups of G because A and B are Hall subgroups. Thus by the Sylow theorems there exists $x \in G$:

$x^{-1}P_B x = P_A$. Put $H = <A, x^{-1}Bx>$. Since P_A is normal in both A and $x^{-1}Bx$ because the latter are nilpotent, P_A is normal in H. But A/P_A and $x^{-1}Bx/P_A$ are nilpotent Hall subgroups of H/P_A, and so by induction they are conjugate in H/P_A. Thus A and $x^{-1}Bx$ are conjugate in H, and so A and B are conjugate as asserted. ■

THEOREM 5.4. *Let G be a finite non-modular linear group of degree n. Let π be the set of all primes $> n+1$. Then G has an abelian Hall π-group, and moreover all the Hall π-groups of G are conjugate in G.*

Proof. By Theorem 5.1 it is enough to consider the case where $G \subseteq GL\ n, \mathbf{C})$. Also from Corollary 5.2 all π-subgroups of G are abelian; in particular, Lemma 5.4 then shows that all Hall π-groups of G are conjugate. Thus it remains to show that G has a Hall π-group, and we shall use induction on the order g of G.

Case (a) $(G \neq G')$. In this case there exists a normal subgroup M of G with $M \supseteq G'$ and so $|G : M| = q$ is prime. By induction M has a Hall π-group H. If $q \notin \pi$, then H is a Hall π-group of G as required. If $q \in \pi$, then choose a q-element $x \in G$ such that $G = <x, M>$. Since the π-subgroups of G are abelian, Lemma 5.4 shows that all Hall π-groups of M are conjugate to H in M; thus there are $|M{:}N_M(H)| = h$, say, such groups. Now x permutes this set of Hall π-groups by conjugation. Since $q \in \pi$, $q \nmid h$. Since x is a q-element, we conclude that for some Hall π-group K, say, of M we have $x^{-1}Kx = K$; that is $\{K\}$ is an orbit under the action of x. It is now clear that $<K, x>$ is the required Hall π-group in this case.

Case (b) $(G = G')$. Since G is completely reducible by Theorem 2.3, we may suppose by induction on degree that G is irreducible. Since $G' = G$ and $\det(x^{-1}y^{-1}xy) = 1$ for all $x, y \in G$, we have

$G \subsetneq SL(n, \mathbf{C})$. Thus by Schur's lemma (§2.6), $Z(G) \subseteq \{\zeta 1 | \zeta \in \mathbf{C}, \det(\zeta 1) = \zeta^n = 1\}$. Therefore, if $x \neq 1$ is a π-element of G then $C_G(x) \neq G$.

Now it follows from Lemma 5.3C that if $p_i \in \pi$ divides $g = |G|$, then for some $x_i \in G$, $\chi(x_i)$ requires p_ith roots of 1. Thus if $p_1, \ldots, p_s$ are the different primes in π dividing g, G has an element x of order $p_1 \ldots p_s$ by Lemma 5.3B. As we have just seen $C_G(x) \neq G$ and so by induction $C_G(x)$ has a Hall π-group H. We shall conclude by showing that H is a Hall π-group of G. For each p_i $(1 \leqslant i \leqslant s)$, p_i divides the order of x and hence of $C_G(x)$; therefore $p_i | \, |H|$. Choose P_i as a Sylow p_i-group of G such that $P_i \cap H \neq 1$ is a Sylow p_i-group of H (using the Sylow theorems). But $<P_i, H> \subsetneq C_G(P_i \cap H) \neq G$, and so by induction H is a Hall π-group in $C_G(P_i \cap H)$; hence $|P_i|$ divides $|H|$. Since this is true for $i = 1, \ldots, s$, we conclude that H is a Hall π-group of G. ■

Exercise. Show that the subgroup G of $GL(n, \mathbf{C})$ generated by

$$\begin{bmatrix} i & 0 \\ 0 & i \end{bmatrix}, \quad \frac{1}{2}\begin{bmatrix} -1+i & -1+i \\ 1+i & -1-i \end{bmatrix} \quad \text{and} \quad \frac{1}{4}\begin{bmatrix} 2i & 1-i-(1+i)\sqrt{5} \\ -1-i+(1-i)\sqrt{5} & -2i \end{bmatrix}$$

has no Hall π-group where π is the set of primes > 2.

§5.5 The theorem of this section is again due to Blichfeldt. It gives a criterion for a Sylow p-group of a linear group to be normal. The bound has recently been much improved (see §5.8).

Let D be a ring of algebraic integers, and let p be a rational prime. Recall that: $\alpha \equiv \beta \pmod{p} \overset{\text{def}}{\Longleftrightarrow}$ there exists $\gamma \in D$ such that $\alpha - \beta = \gamma p$. Because $p \,|\, \binom{p}{i}$ for $1 \leqslant i \leqslant p-1$, $(\alpha + \beta)^p \equiv \alpha^p + \beta^p$ (mod p) by the binomial theorem, and so by induction we find that for any $\alpha_1, \ldots, \alpha_s \in D$:

$$(\alpha_1 + \dots + \alpha_s)^{p^r} \equiv \alpha_1^{p^r} + \dots + \alpha_s^{p^r} \pmod p \text{ for } r = 1, 2, \dots \tag{5.5.1}$$

Also note that if ξ is an mth root of 1 then

$$\sum_{k=1}^{m} \xi^k = \begin{cases} \xi(\xi^m-1)/(\xi-1) = 0 & \text{if } \xi \neq 1 \\ m & \text{if } \xi = 1 \end{cases} \tag{5.5.2}$$

LEMMA 5.5. *Let p be a prime and m an integer > 0 with $p \nmid m$. Let η be a primitive mth root of 1 in $\mathbf{C}$, and for each prime $q_j | m$ choose a primitive q_jth root of unity ω_j ($j = 1, \dots, s$). Suppose that for certain rational integers $a_0, \dots, a_{m-1}$ we have*

$$\sum_{i=0}^{m-1} a_i \eta^i \equiv 0 \pmod p) \text{ in } \mathbf{Z}[\eta]$$

that is, the left-hand side equals $pf(\eta)$ where $f(X) \in \mathbf{Z}[X]$). Then

$$\sum_{i=0}^{m-1} a_i \epsilon_i \equiv 0 \pmod p$$

where

$$\epsilon_i = \begin{cases} (-1)^t \text{ if } \eta^i = \omega_{j_1} \dots \omega_{j_t} & \text{with } 1 \leqslant j_1 < \dots < j_t \leqslant s \\ 0 & \textit{otherwise.} \end{cases}$$

Proof. For each integer k, $1 \leqslant k \leqslant m$ with $(k, m) = 1$ there is a ring automorphism of $\mathbf{Z}[\eta]$ which sends $\eta \mapsto \eta^k$ and leaves each element of $\mathbf{Z}$ fixed. Thus the hypothesis implies

$$\sum_{i=0}^{m-1} a_i \eta^{ik} \equiv 0 \pmod p \tag{5.5.3}$$

for all k, $1 \leqslant k \leqslant m$, with $(k, m) = 1$. But

$$\sum_{\substack{k=1\\(k,m)=1}}^{m} (1-\omega_1^{-k})\ldots(1-\omega_s^{-k})\,\eta^{ik}$$

$$= \sum_{k=1}^{m} (1-\omega_1^{-k})\ldots(1-\omega_s^{-k})\,\eta^{ik}$$

$$= \sum_{1 \leqslant j_1 < \ldots < j_t \leqslant s} (-1)^t \sum_{k=1}^{m} (\omega_{j_1} \ldots \omega_{j_t})^{-k}\,\eta^{ik}$$

$$= m\,\epsilon_i$$

because by (5.5.2) the inner sum vanishes except when $\eta^i = \omega_{j_1} \ldots \omega_{j_t}$. Since $p \nmid m$, the result now follows by summing the congruences (5.5.3) with weighting factors $(1-\omega_1^{-k})\ldots(1-\omega_s^{-k})$, observing that $\omega_j^{-1} \in \mathbf{Z}[\eta]$ because η is a primitive mth root of 1. ∎

THEOREM 5.5. *Let G be a finite non-modular linear group of degree n and order g. Then for each prime $p|g$ for which $p > (2n+1)(n-1)$, G has a unique (normal) Sylow p-group.*

Proof. By Theorem 5.1 we may suppose $G \subseteq GL(n, \mathbf{C})$. Write $g = p^r m$ where $p \nmid m$, and let ζ, λ and η be primitive gth, p^rth and mth roots of 1 in $\mathbf{C}$, respectively. Put $\ell = p^r$. In order to prove the theorem it is enough to show that for any two p-elements x, y in G, xy is also a p-element; for then the p-elements of G form the unique Sylow p-group.

Let $f(X)$ be the minimal polynomial for x where $f(X)$ has leading coefficient 1 and degree d, say. Since all eigenvalues of x are p^rth roots of 1, the coefficients of $f(X)$ are in $\mathbf{Z}[\lambda]$. Thus for each integer $h \geqslant 0$ there are unique $a_{hi} \in \mathbf{Z}[\lambda]$ such that

$$X^h \equiv \sum_{i=0}^{d-1} a_{hi}(X-1)^i \pmod{f(X)}. \qquad (5.5.4)$$

Then

$$x^h = \sum_{i=0}^{d-1} a_{hi}(x-1)^i$$

and so

$$\mathrm{tr}(x^h y) = \sum_{i=0}^{d-1} a_{hi}\ \mathrm{tr}\{(x-1)^i y\} \quad \text{for } h = 0, 1, \ldots .$$

Now

$$\mathrm{tr}\{(x-1)^i y\} = \sum_{j=1}^{i} \tbinom{i}{j} (-1)^{i-j}\ \mathrm{tr}(x^j y) \in \mathbf{Z}[\zeta]$$

because the eigenvalues of all elements of G are powers of ζ. Thus it follows from (5.5.1) with $\ell = p^r$ that the sum τ_h of the ℓth powers of the eigenvalues of $x^h y$ satisfies

$$\tau_h \equiv \{\mathrm{tr}(x^h y)\}^\ell \equiv \sum_{i=0}^{d-1} a_{hi}^\ell\ \beta_i \pmod p \qquad (5.5.5)$$

where τ_h, $\beta_i \in \mathbf{Z}[\eta]$ and $\beta_i \equiv \{\mathrm{tr}(x-1)^i y\}^\ell \pmod p$.

We now show that $a_{hi}^\ell \equiv \binom{h}{i} \pmod p$. Indeed, the roots of $f(X)$ are all powers of λ and $\lambda^\ell = 1$, so $f(X)^\ell \equiv (X^\ell - 1)^d \pmod p$ by (5.5.1). Thus the ideal J in $\mathbf{Z}[\zeta, X]$ generated by $f(X)^\ell$ and p can also be generated by $(X^\ell - 1)^d$ and p. But from (5.5.4) and (5.5.1) we have

$$\sum_{i=0}^{d-1} a_{hi}^\ell (X^\ell - 1)^i \equiv X^{h\ell} = (1 + X^\ell - 1)^h$$

$$\equiv \sum_{i=0}^{d-1} \binom{h}{i} (X^\ell - 1)^i \pmod J.$$

Therefore $a_{hi}^\ell \equiv \binom{h}{i} \pmod p$ as asserted.

If $0 \leqslant i < p$, then $\binom{h}{i} = h(h-1)\ldots(h-i+1)/i!$ is congruent (mod p) to a polynomial $a_i(h)$ in h with integral coefficients and of degree i. Then (5.5.5) shows

$$\tau_h \equiv \sum_{i=0}^{d-1} a_i(h)\,\beta_i \pmod{p}, \tag{5.5.6}$$

where each $a_i(h)$ is an integer polynomial of degree $\leqslant d-1$ in h.

Now suppose that, contrary to the assertion of the theorem, xy is not a p-element. Applying Lemma 5.5 to (5.5.6) we find t_h, $b_i \in \mathbf{Z}$ such that

$$t_h \equiv \sum_{i=0}^{d-1} a_i(h) b_i \pmod{p} \tag{5.5.7}$$

with $|t_h| \leqslant n$ because τ_h is a sum of n roots of unity. Moreover, since xy is not a p-element, we may choose $\omega_1, \ldots, \omega_s$ in the lemma so that $t_1 \neq n$. The right-hand side of (5.5.7) is then a polynomial of degree $\leqslant d-1$ and it is not a constant polynomial because $t_1 \not\equiv t_0 = n \pmod{p}$. Since p is prime, at most $d-1$ different values of h (mod p) yield the same value (mod p) for this polynomial. Thus for $0 \leqslant h < p$, the right-hand side runs over at least $p/(d-1)$ different values (mod p). On the other hand, $-n \leqslant t_h \leqslant n$, and so the left-hand side of (5.5.7) takes at most $2n+1$ values. Hence $2n+1 \geqslant p/(d-1)$. Thus $p \leqslant (2n+1)(d-1) \leqslant (2n+1)(n-1)$ contrary to the hypothesis. Therefore we conclude xy is a p-element and the theorem follows. ■

Exercises. 1. Making a careful analysis of (5.5.7) show that when $n = 3$ the conclusion of theorem holds if $p > 7$.

2. Let G be a finite non-modular linear group of degree n.

(a) If for some prime p, G has an element of order p^k where

$p^{k-1} \geqslant n$, show that G has a non-trivial normal p-subgroup. [*Hint:* Use Theorem 3.6B.]

(b) Show that the condition in (a) is satisfied if G has a Sylow p-group of order p^e where $e \geqslant n\,\{\log n/\log p + 1/(p-1)\}$. [*Hint:* Use Exercise 2 of §4.6.]

§5.6 The following result is due to A. Loewy (1896) and E. H. Moore (1898).

We begin by recalling a few facts about positive definite Hermitian matrices. A matrix $x \in M(n, \mathbf{C})$ is *Hermitian* if it equals its complex conjugate transpose x^*; it is *positive definite* if it is Hermitian and for each n-vector $v \neq 0$ ($1 \times n$ matrix over $\mathbf{C}$) the 1×1 real matrix $v\,x\,v^* > 0$. A matrix $x \in M(n, \mathbf{C})$ is *unitary* if $x^*x = xx^* = 1$. The following facts are proved in any good book on linear algebra.

(a) The sum of a finite number of $n \times n$ positive definite matrices is positive definite.

(b) $x \in M(n, \mathbf{C})$ is positive definite $\Longleftrightarrow$ $x = uu^*$ for some non-singular $u \in M(n, \mathbf{C})$.

THEOREM 5.6. *For any finite subgroup G of $GL(n, \mathbf{C})$ there exists $u \in GL(n, \mathbf{C})$ such that $u^{-1}Gu$ is a group of unitary matrices.*

Proof. Define $w = \sum_{y \in G} yy^*$. By (a) and (b) above, w is positive definite and so $w = uu^*$ for some $u \in GL(n, \mathbf{C})$ by (b). However, for each $x \in G$, $xwx^* = \sum_{y \in G} (xy)(xy)^* = w$, because xy runs through G as y does. Hence $xuu^*x^* = uu^*$ for all $x \in G$; or equivalently $(u^{-1}xu)(u^{-1}xu)^* = 1$. Thus $u^{-1}Gu$ is a group of unitary matrices. ■

§5.7 The theorem of this section is due to C. Jordan (1878) who discovered the result while studying the space groups of crystallography. We give a proof based on one of Schur (1911). This proof uses Theorem 5.6 and some properties of matrix norms which we now discuss.

We define the (Hilbert) *norm* $||x||$ for each $x \in M(n, \mathbf{C})$ as follows. Let $< \cdot, \cdot >$ be the standard inner product on the vector space $\mathbf{C}^n$ of all n-vectors over $\mathbf{C}$, namely $<u, v> = \Sigma\, u_i \bar{v}_i$ if $u = (u_1, \ldots, u_n)$ and $v = (v_1, \ldots, v_n)$ and the bar denotes complex conjugation. Then $||x||$ is the non-negative real number such that

$$||x||^2 = \sup\{<vx, vx> \mid v \in \mathbf{C}^n, <v, v> = 1\}.$$

(Note that the supremum is finite because it is over a compact set.) We have the following properties.

(N1). $||x+y|| \leqslant ||x|| + ||y||$ for all $x, y \in M(n, \mathbf{C})$. Indeed, for all $v \in \mathbf{C}^n$, $<v(x + y), v(x + y)> = <vx, vx> + <vy, vy> + \{<vx, vy> + <vy, vx>\} \leqslant <vx, vx> + <vy, vy> + 2\{<vx, vx> \cdot <vy, vy>\}^{1/2}$ by the Schwarz inequality. Taking the supremum over all $v \in \mathbf{C}^n$, $<v, v> = 1$, we conclude $||x + y||^2 \leqslant ||x||^2 + ||y||^2 + 2||x||\,||y||$ as required.

(N2). $||xy|| \leqslant ||x||\,||y||$. Indeed, for all $v \in \mathbf{C}^n$ with $\lambda_v = <vx, vx> \neq 0$ we define $u_v = \lambda_v^{-1/2}\, vx \in \mathbf{C}^n$ where $<u_v, u_v> = 1$. Then $<vxy, vxy> = \lambda_v <u_v y, u_v y> = <vx, vx><u_v y, u_v y>$. Taking the supremum of the left-hand side over $v \in \mathbf{C}^n$, $<v, v> = 1$, we get $||xy||^2 \leqslant ||x||^2 ||y||^2$ as asserted.

(N3). If $x \in M(n, \mathbf{C})$ is unitary, then $<vx, vx> = <v, v>$ and so $||x|| = 1$. In general, $||x||$ is not less than the largest absolute value of an eigenvalue of x (since v may be an eigenvector), so $|\mathrm{tr}\ x| \leqslant n||x||$ for all $x \in M(n, \mathbf{C})$.

(N4). If $x, y \in M(n, \mathbf{C})$ are unitary, then by (N1)–(N3) we have $||1 - x^{-1}y^{-1}xy|| = ||x^{-1}y^{-1}(yx - xy)|| \leqslant ||x^{-1}||\,||y^{-1}||\,||(1-y)(1-x) - (1-x)(1-y)|| \leqslant 2||1-x||\,||1-y||$.

(N5). If $x = [\xi_{ij}] \in M(n, \mathbf{C})$ and $v = (v_1, \ldots, v_n)$, then $\langle vx, vx\rangle = \sum_j |\sum_i v_i \xi_{ij}|^2 \leqslant n \max |\xi_{ij}|^2 \langle v, v\rangle$. Hence $||x|| < \sqrt{n} \max |\xi_{ij}|$.

As a final preparation for the theorem we prove the following lemma.

LEMMA 5.7. *Let* $x, y, z \in M(n, F)$ *for some field* F. *Suppose* y *is non-singular,* $yxy^{-1} = xz$, *and* $zy = yz$. *Then:*

(1) *Either* $\operatorname{tr} x = 0$ *or* z *has an eigenvalue equal to* 1.

(2) *If* x, y *and* z *are unitary, and* $||1-x|| < 1$, *then* $z = 1$.

Proof. (1) Let $f(X) = \sum_{i=0}^{n} a_i X^i$ be the characteristic polynomial for z. By induction on k, $y^{-k}xy^k = xz^k$, and so $\operatorname{tr} xz^k = \operatorname{tr}(y^{-k}xy^k) = \operatorname{tr} x$ for $k = 0, 1, \ldots$. By the Cayley–Hamilton theorem $f(z) = 0$, and so

$$0 = \operatorname{tr}(xf(z)) = \sum_{i=0}^{n} a_i \operatorname{tr} xz^i$$

$$= \sum_{i=0}^{n} a_i \operatorname{tr} x = f(1) \cdot \operatorname{tr} x.$$

Thus either $\operatorname{tr} x = 0$ or $f(1) = 0$; in the latter case z has 1 as an eigenvalue.

(3) Since z is unitary, there is a unitary matrix u such that for some r, $0 \leqslant r \leqslant n$,

$$u^{-1}zu = \begin{bmatrix} 1 & 0 \\ 0 & z_2 \end{bmatrix}$$

where $1 = 1_r$ is an $r \times r$ block and z_2 is an $(n-r) \times (n-r)$ diagonal block with no eigenvalue 1. We must prove $r = n$. Suppose the contrary. Since $zy = yz$ we have

$$u^{-1}yu = \begin{bmatrix} y_1 & 0 \\ 0 & y_2 \end{bmatrix} \quad \text{and} \quad u^{-1}xu = \begin{bmatrix} x_1 & x_3 \\ x_4 & x_2 \end{bmatrix}$$

where the matrices are partitioned in a similar way to $u^{-1}zu$. Since $||1-x|| < 1$, $||1-u^{-1}xu|| \leqslant ||u^{-1}|| \cdot ||1-x|| \cdot ||u|| < 1$ by (N2) and (N3) above. Moreover, for each $v \in \mathbf{C}^{n-r}$ with $\langle v, v\rangle = 1$, the augmented vector $v' = (0\ v) \in \mathbf{C}^n$ satisfies $\langle v', v'\rangle = 1$ (here 0 is the zero r-vector). Thus it easily follows that $||1 - x_2|| < 1$, and then (N3) shows that $|\mathrm{tr}\ x_2| \geqslant |\mathrm{tr}\ 1| - |\mathrm{tr}(1 - x_2)| \geqslant (n-r) - (n-r)||1 - x_2|| > 0$. But from $xz = y^{-1}xy$ we find $x_2z_2 = y_2^{-1}x_2y_2$ and we also have $z_2y_2 = y_2z_2$. Since $\mathrm{tr}\ x_2 \neq 0$, z_2 has 1 as an eigenvalue by (1). This contradicts our hypothesis on z_2. Thus we conclude $r = n$ and $z = 1$. ■

THEOREM 5.7. *Let G be a finite non-modular linear group of degree n. Then G has a normal abelian subgroup A with index* $|G : A| \leqslant (49n)^{n^2}$.

Proof. From Theorems 5.1 and 5.6 it is enough to consider the case where G is a group of unitary matrices in $GL(n, \mathbf{C})$. We define A as the subgroup generated by all $x \in G$ with $||1 - x|| < 1/2$. From (N2) and (N3) above, we have $||1 - y^{-1}xy|| \leqslant ||y^{-1}||\ ||1-x||\ ||y|| = ||1-x||$ for all $x, y \in G$ and so A is normal in G. It remains to show that A is abelian and $|G : A| \leqslant (49n)^{n^2}$.

(a) *A is abelian.* It is enough to show that $||1 - x|| < 1/2$ and $||1 - y|| < 1/2$ imply $x^{-1}y^{-1}xy = 1$. Define $z_k \in G$ by $z_0 = x$ and $z_k = z_{k-1}^{-1}\ y^{-1}\ z_{k-1}\ y$ $(k = 1, 2, \ldots)$. By (N4) we have

$$||1 - z_k|| \leqslant 2||1 - z_{k-1}||\ ||1 - y|| < ||1 - z_{k-1}|| \qquad (5.7.1)$$

unless $||1 - z_{k-1}|| = 0$. Since G is finite, we conclude that

$||1 - z_m|| = 0$ for some m, and then $z_m = 1$. Choose m as the smallest index such that $z_m = 1$; we have to prove $m = 1$. If $m > 1$, then y commutes with $z_{m-1} = z_{m-2}^{-1}y^{-1}z_{m-2}y$. But $||1 - z_{m-2}|| \leqslant ||1 - x|| < 1/2$ by (5.7.1) and so $z_{m-1} = 1$ by Lemma 5.7(2). This contradicts the choice of m. Hence $m = 1$, $x^{-1}y^{-1}xy = 1$, and we conclude A is abelian.

(b) *The index* $|G : A| \leqslant (49n)^{n^2}$. Equivalently, we shall show that if $m > (49n)^{n^2}$, and $x_1, \ldots, x_m \in G$, then $Ax_i = Ax_j$ for some $i \neq j$. Since each $x \in G$ is unitary all its entries have absolute value $\leqslant 1$, and so there is a mapping $x \mapsto P_x$ of G into the $2n^2$-dimensional hypercube $[-1, 1]^{2n^2}$ where the cooredinates of P_x are the real and imaginary parts of the n^2 entries of x in some prescribed order. This hypercube is a union of $([6\sqrt{n} + 1])^{2n^2} \leqslant (49n)^{n^2}$ smaller hypercubes each of side length $1/3\sqrt{n}$. Since $m > (49n)^{n^2}$, there are x_i, x_j $(i \neq j)$ such that P_{x_i}, P_{x_j} lie in the same small hypercube. This means that each entry of $x_i - x_j$ has absolute value $\leqslant \{(3\sqrt{n})^{-2} + (3\sqrt{n})^{-2}\}^{1/2} = \sqrt{2}/3\sqrt{n}$. Hence by (N2), (N3) and (N5) above we have $||1 - x_jx_i^{-1}|| \leqslant ||x_i - x_j||\,||x_i^{-1}|| = ||x_i - x_j|| \leqslant \sqrt{n}\,\sqrt{2}/3\,\sqrt{n} < 1/2$. Thus $x_jx_i^{-1} \in A$ and $Ax_i = Ax_j$ as required. ■

Exercises. 1. Let G be a finite non-modular linear group of degree n. Show that if G has an abelian subgroup $B \supseteq Z(G)$ with $|B : Z(G)| > (4\pi)^n$, then G has a normal abelian subgroup A with $B \supseteq A \supset Z(G)$. In particular, if G is irreducible and the underlying field is algebraically closed, then G is imprimitive by Corollary 4.2B. [*Hint:* We may suppose G consists of unitary matrices, and the elements of B are diagonal.]

2. Let G be a finite non-modular group of degree n. If G has a Sylow p-group of order p^e where $e > n\{\log 4\pi/\log p + 1/(p-1)\}$, then

G has a non-trivial normal p-subgroup. (This improves Exercise 2 of §5.5 when $n \geqslant 13$.)

3. Let F be a field of characteristic $p > 0$ which is the algebraic closure of its prime subfield. Let $n \geqslant 2$. Show that for each constant $\beta > 0$, there is a finite subgroup of $GL(n, F)$ which has no abelian normal subgroup of index $< \beta$.
[*Hint:* let E be a finite subfield of F and consider $GL(n, E)$.]

§5.8 *Notes and References*

Theorem 5.2 was first proved in

Itô, N. On the degrees of irreducible representations of a finite group, *Nagoya Math. J.* **3**, 5–6 (1961).

It has been generalized by W. F. Reynolds who showed that if G is a finite irreducible subgroup of $GL(n, \mathbf{C})$, and H is a normal subgroup of G, then n divides $|G : H|d$ where d is the degree of some irreducible representation of H over $\mathbf{C}$; see §56 of the book of Curtis and Reiner. Another kind of generalization is given in

Dade, E. C. Degrees of modular irreducible representations of p-solvable groups, *Math. Z.* **104**, 141–143 (1968).

The result of Exercise 2 of §5.2 appears in Gorenstein's *Finite Groups*.

The material of §5.3 includes a result from

Brauer, R. A note on theorems of Burnside and Blichfeldt, *Proc. Amer. Math. Soc.* **15**, 31–34 (1964).

Exercise 2 of §5.3 is also due to Brauer, and Exercises 3 and 4 are from

Schur, I. Uber eine Klasse von endlichen Gruppen Linearer Substitutionen, *Sitzber. Preuss. Akad. Wiss.* 77–91 (1905).

The material of §5.5 is essentially a cleaned-up version of Blichfeldt's original proof. He gives an exposition of his proof in

Blichfeldt, H. F. *Finite Collineation Groups*. University of Chicago Press (1917).

A more generally available source is

Miller, G. A., Blichfeldt, H. F. and Dickson, L. E. *Theory and Applications of Finite Groups* (reprint). Dover, New York (1961).

By using results of Brauer, Feit and Thompson have improved Theorem 5.5 to show that the Sylow p-group is normal if $p > 2n + 1$. See

Feit, W. and Thompson J. On groups which have a faithful representation of degree less than $(p-1)/2$, *Pacif. J. Math.* **11**, 1257–1262 (1961).

There have been further more recent results by Feit. We shall later prove analogous theorems (with improved bounds) for solvable and p-solvable groups.

Theorem 5.6 is one of the oldest results in the theory of linear groups. There is a stronger form which implies that any compact subgroup of $GL(n, \mathbf{C})$ (in the usual topology) is conjugate to a group of unitary matrices. This result and its ramifications may be found in §28 of

Pontrjagin, L. *Topological Groups*. (Princeton Mathematical Series, Vol. 2). Princeton University Press (1939).

Theorem 5.7 has had many proofs. For a slightly different proof and further references, see §36 of Curtis and Reiner's book. It is still an open question as to what order of growth the best bound for the index of the abelian subgroup has. Theorem 5.7 is generalized to periodic groups in Theorem 9.5. Recently it has been shown that a finite linear group G of degree n over a field of characteristic $p > 0$ has a normal abelian subgroup A of index $|G : A| \leqslant \beta$ where β depends only on n, p and the order of the Sylow p-groups of G (see Exercise 3 of §5.7). See

Brauer, R. and Feit, W. An analogue of Jordan's theorem in characteristic p, *Ann. Math.* **84**, 119–131 (1966).

It has been pointed out to me by L. G. Kovacs that this result shows that if F is a field of characteristic $p > 0$, then each finite subgroup G of $GL(n, F)$ is isomorphic to a subgroup of $GL(m, \mathbf{C})$ where m depends only on n, p and the order of the Sylow p-groups of G. [*Proof:* G has a non-modular subgroup H of index m/n by Brauer and Feit's theorem. Apply Theorem 5.1 to H and induce to get a representation of G.] This may be compared with Theorem 5.1.

An interesting question is to classify the finite subgroups of $GL(n, \mathbf{C})$ for small degree n. This was done by Klein (1876) for $n = 2$, by Jordan (1878) for $n = 3$ and by Blichfeldt (1905) for $n = 4$ (see the books by Blichfeldt quoted above). Recent results for $n = 5$, $n = 7$ have been obtained by Brauer and D. B. Wales. See

Wales, D. B. Finite linear groups in seven variables, *Bull. Am. math. Soc.* **74**, 197–198 (1968).

CHAPTER 6

Solvable and Nilpotent Groups

§6.1 Many of the applications of linear groups to the structure theory of abstract groups occur when the groups are solvable (see Exercises 1 and 2 below). Moreover in the case of solvable linear groups we can obtain several precise theorems which are not available in the general case. Although the results of this chapter arose in various contexts we are able to unify many of their proofs by applying Theorem 4.5.

The theorems of §6.2 show that a solvable linear group has its solvable length bounded by a function of its degree, and hence that a locally solvable linear group is always solvable. In §6.3 and §6.4 we deal with a result which sharpens Blichfeldt's Theorem 5.5 for solvable groups, and at the same time we find a bound (in terms of the degree) on the index of a maximal normal abelian subgroup of a completely reducible solvable linear group (compare with Theorem 5.7). In §6.5 we give results of Suprunenko on completely reducible nilpotent groups. Finally, Theorem 6.6 shows how the complete reducibility of a solvable linear group is largely a 'local' matter; this is in contrast to the general case for linear groups (see §2.5).

Exercises. 1. Let G be a finite group and let $G = G_0 \supset G_1 \supset \ldots \supset G_n = 1$ be a principal series for G. Show that $x \in \operatorname{Fit} G$ (see §4.4) $\Longleftrightarrow [x, G_{i-1}] \subseteq G_i$ for $i = 1, 2, \ldots n$.

2. If G is solvable in Exercise 1, then each G_{i-1}/G_i is an elementary abelian p_i-group of order $p_i^{r_i}$, say, for some prime p_i. Show that $G/\mathrm{Fit}(G)$ is isomorphic to a subgroup of the direct product $\bigotimes_{i=1}^{n} GL(r_i, p_i)$.

§6.2 We begin with a theorem of B. Huppert (1958) which strengthened a qualitative result of Zassenhaus (1938); see §6.7. The method of proof given here is typical of the proofs of several theorems of this chapter.

THEOREMS 6.2A. *Let G be a solvable subgroup of $GL(V)$ where V is a vector space of dimension n over a field F. Then the 2nth derived group $G^{(2n)} = 1$; that is, the solvable length ℓ of G is at most $2n$.*

Proof. We may suppose F is algebraically closed, and we proceed by induction on n. If $n = 1$, G is abelian and so $G^{(1)} = 1$; therefore suppose $n > 1$. We have three cases.

Case 1 (G is reducible). If $W \neq 0$ or V is a G-space of V, then the mapping $x \to (x|W, x|V/W)$ is a homomorphism of G into the direct product $G|W \times G|(V/W)$ of solvable linear groups of smaller degrees. The kernel of this homomorphism is readily seen to be abelian. Thus if $r = \dim W$, then induction shows $\ell \leqslant 1 + \max\{2r, 2(n-r)\} \leqslant 2n-1 < 2n$ as required.

Case 2 (G irreducible but imprimitive). Let $V = \bigoplus_{i=1}^{m} V_i$ $(m > 1)$ where the subspaces V_i are permuted under the action of G. By Theorem 4.2B there is a normal subgroup N of G such that G/N is isomorphic to a permutation group of degree n and N is isomorphic to a subgroup of $N|V_1 \times \ldots \times N|V_m$ where each V_i has dimension $d = n/m$. Now a permutation group of degree m is isomorphic to a group of permutation matrices of degree m (see §1.3), and the latter group is always reducible since $m \geqslant 2$ and the space spanned by the vector $(1\,1\,\ldots\,1)$ is an invariant subspace. Hence by Case 1 we we conclude $(G/N)^{(2m-1)} = 1$. On the other hand induction shows that $(N|V_i)^{(2d)} = 1$ for each i, and $(N|V_i)^{(1)} = 1$ if $d = 1$. Thus $N^{(2d)} = 1$, and $N^{(1)} = 1$ if $d = 1$. Therefore

$$\ell \leqslant (2m-1) + 2d < 2n \quad \text{if } d = n/m > 1$$

and

$$\ell \leqslant (2n-1) + 1 = 2n \quad \text{if } d = 1.$$

Thus the result follows in this case.

Case 3 (G is primitive). We use Theorem 4.5 and note from that theorem that $(\text{Fit } G)^{(2)} = 1$. Let $n = p_1^{\ell_1} \ldots p_s^{\ell_s}$ be the canonical prime decomposition of n. Since $d|n$ in Theorem 4.5, the solvable length ℓ of G is at most 2 more than the maximum solvable length of any solvable subgroup of one of the $\text{Sp}(2\ell_i, p_i)$. Since we must have $2^{\ell_i} \leqslant p_i^{\ell_i} \leqslant n$ for all i, we conclude by induction that

$$\ell \leqslant 2 + \max\{2\ell_1, \ldots, 2\ell_s\} \leqslant 2n$$

except in the cases $n = 2$ or $n = 2^2$. In these latter cases $|Sp(2,2)| = 2.3$ and $|Sp(4,2)| = 2^4 \cdot 3^2 \cdot 5$ by Theorem 1.5. Thus when $n = 2$, $\ell \leqslant 2{+}2 = 2n$. On the other hand when $n = 2^2$ we know that $Sp(4,2)$ is not solvable by Theorem 1.6, and so the number of prime factors in the order of any solvable subgroup of $Sp(4,2)$ is at most 6; hence $(G/\text{Fit } G)^{(6)} = 1$, so $\ell \leqslant 2{+}6 = 2n$.

This completes the proof of this case and hence concludes the proof of the theorem. ■

As an immediate application we have the following result.

THEOREM 6.2B. *Let G be a linear group. If G is* locally solvable *(that is, each finitely generated subgroup is solvable), then G is solvable.*

Proof. Suppose that G has degree n. Then for any sequence of 2^{2n} elements $x_1, x_2, \ldots, x_{2^{2n}}$ of G the compound commutator

$$[\ldots [[[x_1, x_2], [x_3, x_4]], [[x_5, x_6], [x_7, x_8]]], \ldots]$$

is contained in $H^{(2n)}$ where H is the subgroup generated by the x_i. By hypothesis H is solvable, so $H^{(2n)} = 1$ by Theorem 6.2A. Thus all such commutators equal 1. But the set of all such commutators generates $G^{(2n)}$, and so $G^{(2n)} = 1$. ■

Exercises. 1. Show that the subgroup of $GL(2, \mathbf{C})$ generated by

$$x = \frac{1}{\sqrt{2}} \begin{bmatrix} -\zeta & -\zeta \\ \zeta^{-1} & -\zeta^{-1} \end{bmatrix} \quad \text{and} \quad y = \begin{bmatrix} \zeta^{-1} & 0 \\ 0 & \zeta \end{bmatrix}$$

where ζ is a primitive 8th root of unity, has order 48 and solvable length 4.

2. Let G be a completely reducible linear group of degree n and order g. If G is solvable, and each prime $p|g$ is odd and satisfies $p(2p-1) > n$, show that $G^{(3)} = 1$. (Compare with Corollary 5.2. Note that the condition that G is solvable is implied by the condition that g is odd from the renowned Feit–Thompson theorem.)

3. Let G_n be the subgroup of $GL(2, \mathbf{C})$ generated by

$$\begin{bmatrix} 0 & -1 \\ 1 & 0 \end{bmatrix} \quad \text{and} \quad \begin{bmatrix} \zeta & 0 \\ 0 & \zeta^{-1} \end{bmatrix}$$

where ζ is a primitive 2^nth root of unity. Show that G_n is a finite

2-group which is nilpotent of class n ($n \geqslant 1$), and that $G = \bigcup_{n=1}^{\infty} G_n$ is an infinite locally nilpotent 2-group which is not nilpotent; indeed the center of $G/Z(G)$ is trivial.

4. Let $R_m = \mathbf{Z}/(p^m)$ be the ring of integers (mod p^m) where p is prime. Let $G = \{1+pu \mid u \in M(n,R_m)\} \subseteq GL(n,R_m)$. Show that G is a solvable group of solvable length k if $m = 2^{k-1}$. (Thus Theorem 6.2A does not generalize to groups over commutative rings).

5. Complete the last part of the proof of Theorem 6.2A without appealing to Theorem 1.6.

§6.3 Itô (1953) showed that a finite solvable subgroup of $GL(n, \mathbf{C})$ has a normal Sylow p-group whenever $p > n+1$ (and $p > n$ is sufficient if $p-1$ is not a power of 2). We shall give a generalization of this theorem in this section.

Let G be a finite group. For each prime p we define the *p-core* K of G as the (unique) largest normal p-subgroup of G. If P is a Sylow p-group of G, it is readily seen that $K = P \cap \text{Fit}\, G$; thus P/K is isomorphic to a Sylow p-group of $G/\text{Fit}\, G$ (Fit G is the Fitting subgroup of G; see §4.4).

The proof of the following lemma is left as an exercise.

LEMMA 6.3. *If N is a normal subgroup of a group G, then* $\text{Fit}\, N \subseteq \text{Fit}\, G$ *and so* $|G : \text{Fit}\, G|$ *divides* $|G : N|\,|N : \text{Fit}\, N|$. *If H is a subgroup of a direct product* $H_1 \times \ldots \times H_m$, *then* $\text{Fit}\, H \supseteq H \cap \text{Fit}(H_1 \times \ldots \times H_m)$ *and so* $|H : \text{Fit}\, H|$ *divides*

$$|H_1 \times \ldots \times H_m : \text{Fit}(H_1 \times \ldots \times H_m)| = \prod_{i=1}^{m} |H_i : \text{Fit}\, H_i|. \blacksquare$$

THEOREM 6.3. *Let G be a solvable completely reducible linear group of degree n. Then* $|G : \text{Fit}\, G|$ *is finite and divides* $\prod_p p^{\lambda_p(p)}$

(where the product is over all primes $p \leqslant n+1$) where

$$\lambda_p(n) = \begin{cases} \sum_{i=0}^{\infty} [n/p'p^i] & \text{if } p \text{ is odd} \\ [4n/3]-1 & \text{if } p = 2. \end{cases} \tag{6.3.1}$$

Here $p' = p-1$ or p depending on whether p is a Fermat prime (that is, of the form 2^k+1) or not; $[x]$ denotes the largest integer $\leqslant x$. Note that $\lambda_p(n) = 0$ when $p > n+1$, and that the formally infinite sum in (6.3.1) has only a finite number of non-zero terms.

From the observation at the beginning of this section, we can conclude further that, if G is finite and K and P are the p-core and a Sylow p-group of G, respectively, then $|P : K| < p^{\lambda_p(n)}$. In particular if $p' > n$, then $P = K$ and so $P \lhd G$ (Itô's theorem).

Proof. Let p be a fixed prime and put $\lambda(n)$ for $\lambda_p(n)$ and $\lambda(G)$ for the exponent of p in $|G : \text{Fit } G|$ when the latter is finite. We have to prove $\lambda(G) \leqslant \lambda(n)$. By Corollary 2.8C it is enough to consider the case where G is a completely reducible subgroup of $GL(V)$ and V is a vector space of dimension n over an algebraically closed field. We proceed by induction on n; the result is true if $n = 1$, so suppose $n > 1$.

Case 1 (G is reducible). Then $V = W_1 \oplus W_2$ where the W_i are non-trivial G-spaces. The mapping $x \mapsto (x|W_1, x|W_2)$ is an isomorphism of G onto a subgroup of $G|W_1 \times G|W_2$. Hence by induction and Lemma 6.3 we conclude $|G : \text{Fit } G| < \infty$ and $\lambda(G) \leqslant \lambda(G|W_1) + \lambda(G|W_2)$. If $\dim W_1 = r$, this shows $\lambda(G) \leqslant \lambda(r) + \lambda(n-r) \leqslant \lambda(n)$ as required.

Case 2 (G is irreducible but imprimitive). Let $V = \bigoplus_{i=1}^{m} V_i$ $(m > 1)$ where the subspaces V_i are permuted under the action of G. Let N be the normal subgroup of G consisting of all $x \in G$ such that

$V_i x = V_i$ ($i = 1, \ldots, m$). Then N is isomorphic to a subgroup of $N|V_1 \times \ldots \times N|V_m$ and G/N is isomorphic to a subgroup of the symmetric group S_m of order $m!$ (Theorem 4.2B). Now $N|V_i$ is a completely reducible group of degree $d = n/m$ by Theorem 4.2B, and so by induction $\lambda(N|V_i) \leqslant \lambda(d)$. Therefore from Lemma 6.3 we conclude

$$\lambda(G) \leqslant e + \lambda(N) \leqslant e + m\lambda(d) \tag{6.3.2}$$

where $e = \sum_{i=1}^{\infty} [m/p^i]$ is the exponent to which p divides $m!$. If $p = 2$, (6.3.2) shows $\lambda(G) \leqslant \lambda(n)$ immediately. If p is odd, then choose the integer j so that $p^j p' \leqslant d < p^{j+1} p'$; then (6.3.2) gives

$$\begin{aligned}\lambda(G) &\leqslant e + m\lambda(d) \\ &\leqslant \sum_{i=1}^{\infty} [md/p^{j+i}p'] + \sum_{i=0}^{j} [md/p^i p'] \\ &= \lambda(md) = \lambda(n).\end{aligned}$$

Thus the result is proved in this case.

Case 3 (G is primitive). We shall apply Theorem 4.5. Let $n = p_1^{\ell_1} \ldots p_s^{\ell_s}$ be the prime decomposition of n. By Theorem 4.5, $\lambda(G) \leqslant \nu_1 + \ldots + \nu_s$ where p^{ν_i} is the order of a Sylow p-group in $\mathrm{Sp}(2\ell_i, p_i)$. If we can prove that $\nu_i \leqslant \lambda(p_i^{\ell_i})$ for each i, then it readily follows that

$$\lambda(G) \leqslant \sum_{i=1}^{s} \nu_i \leqslant \sum_{i=1}^{s} \lambda(p_i^{\ell_i}) \leqslant \lambda(n).$$

Thus it remains to show that if q is a prime and $\ell \geqslant 1$ is an integer, then the exponent ν of p in $|\mathrm{Sp}(2\ell, q)|$ is at most $\lambda(q^{\ell})$. This we proceed to do.

First by, $|Sp(2\ell,q)| = q^{\ell^2} \prod_{i=1}^{\ell} (q^{2i}-1)$ by Theorem 1.5. Therefore if $p = q$, then $\nu = \ell^2 \leqslant p^{\ell-1} + p^{\ell-2} + \ldots + 1 \leqslant \lambda(p^\ell)$ if p is odd, and $\nu = \ell^2 \leqslant [2^{\ell+2}/3]-1 = \lambda(2^\ell)$ if $p = 2$. Thus the result holds in this case. On the other hand, if $p \neq q$, then p^ν divides $(q-1)(q+1) \ldots (q^\ell-1)(q^\ell+1)$, and since $q^i-1 > q^{i-1}+1$ unless $q^i-1 = 3$ we conclude that p^ν divides $3(q^\ell+1)!$ Thus:

(a) If $p \geqslant 5$, then

$$\nu \leqslant \sum_{i=1}^{\infty} [(q^\ell+1)/p^i] \leqslant \sum_{i=0}^{\infty} [q^\ell/p'p^i] = \lambda(q^\ell).$$

(b) If $p = 3$, then

$$\nu \leqslant 1 + \sum_{i=1}^{\infty} [(q^\ell+1)/3^i];$$

hence if $q^\ell \geqslant 8$,

$$\nu \leqslant \sum_{i=0}^{\infty} [q^\ell/3^i 2] = \lambda(q^\ell),$$

whilst $\nu \leqslant \lambda(q^\ell)$ directly if $q^\ell = 2,4,5$ or 7.

(c) If $p = 2$,

$$\nu \leqslant \sum_{i=1}^{\infty} [(q^\ell+1)/2^i] \leqslant q^\ell \leqslant \lambda(q^\ell)$$

because $q^\ell \geqslant 3$.

Thus $\nu \leqslant \lambda(q^\ell)$ has been verified in all cases and the theorem is proved. ■

COROLLARY 6.3. *Let G be a finite solvable linear group of degree n over a field of characteristic $p > 0$ (G is not assumed completely reducible). Then with the notation of the Theorem,* $|P:K| < p^{\lambda_p(n)}$.

Proof. From Corollary 2.4 and the Theorem. ∎

§6.4 The theorem of this section is due to Mal'cev (1948). It may be compared with Theorem 5.7.

THEOREM 6.4. *There exists a bound $\beta(n)$ depending only on n such that if G is a complete reducible solvable linear group of degree n, then G has a normal abelian subgroup A of index* $|G:A| \leqslant \beta(n)$.

Proof. Let $\alpha(n) = \prod_p p^{\lambda_p(n)}$ (with the product over all primes $p \leqslant n+1$) where $\lambda_p(n)$ is defined in (6.3.1). Then $|G:\mathrm{Fit}\ G| \leqslant \alpha(n)$ by Theorem 6.3. On the other hand, Fit G is a locally nilpotent group (see §4.4), and Fit G is completely reducible by Theorem 2.2. Thus, by Corollary 4.6B, Fit G is a monomial group, and so it has a normal abelian subgroup B such that (Fit $G)/B$ is isomorphic to a permutation group of degree n (see §4.6). Thus B is an abelian subgroup of G with index $|G:B| \leqslant \alpha(n)\cdot n!$ In general B is not normal in G, but since Fit $G \subseteq N_G(B)$, B has at most $\alpha(n)$ conjugates. Then, if A is the intersection of these conjugates, A is normal in G and $|G:A| \leqslant \beta(n) = (\alpha(n)n!)^{\alpha(n)}$. ∎

COROLLARY 6.4. *Let G be a solvable linear group of degree n over an algebraically closed field F. Then G has a normal subgroup H of index $\leqslant \beta(n)$ such that H corresponds to a subgroup of $TL(n,F)$ over a suitable basis for the underlying vector space. In particular, H' is a nilpotent group consisting of unipotent elements and H' is nilpotent of class $\leqslant n-1$.*

Proof. Using Theorem 2.4 we find that G has a normal subgroup N such that G/N is isomorphic to a completely reducible group G^ρ of degree n over F and N corresponds to a subgroup of $STL(n,F)$ over a suitable basis. By the Theorem G^ρ has a normal abelian subgroup A of index $\leqslant \beta(n)$; let $H \subseteq G$ be the inverse image under ρ of A. Since A is completely reducible by Theorem 2.2, it correspond to a subgroup of $\mathrm{Diag}(n,F)$ over a suitable basis (Corollary 2.6). It readily follows that over this same basis H corresponds to a subgrou of $TL(n,F)$. Hence H is the required subgroup of G. The final assertion of the Corollary comes from Theorem 2.4 because $H' \subseteq N$. ∎

Note. The value for $\beta(n)$ given in the proof of Theorem 6.4 is very rough. For example, see my paper on Fitting subgroups listed in §6.7.

Exercise. Let G be a locally nilpotent linear group of degree n. Show that G has a normal subgroup H of index $\leqslant \beta(n)$ (defined in Theorem 6.4) such that H is nilpotent of class $\leqslant n-1$.
[*Hint:* Show that a locally nilpotent subgroup of $TL(n,F)$ is nilpotent of class $\leqslant n-1$.]

§6.5 In this section we shall study nilpotent groups. We have already seen (Exercise 3 of §6.2) that the 'nilpotent analogue' of Theorem 6.2 is false. The theorem of the present section is a result of D. Suprunenko (1954) on the index of the center and derived group of a nilpotent linear group.

We shall require the following general lemma.

LEMMA 6.5. *Let G be a group whose center Z has exponent e. If $1 = Z_0 \subseteq Z_1 \subseteq Z_2 \subseteq \ldots$ is the upper central series for G, then $x^e \in Z_i$ for all $x \in Z_{i+1}$ for $i = 0, 1, \ldots$. In particular, if G is nilpotent of class k (so $Z_k = G$), then $x^{e^k} = 1$ for all $x \in G$.*

Proof. We proceed by induction on i to show that $x^e \in Z_i$ for all $x \in Z_{i+1}$. This is true for $i = 0$ by hypothesis so suppose $i > 0$. Since $Z_i/Z_{i-1} = Z(G/Z_{i-1})$, and $[x,y] \in Z_i$ for all $x \in Z_{i+1}$, $y \in G$, we have $x[x,y]Z_{i-1} = [x,y]xZ_{i-1}$. Hence $x(y^{-1}xy)Z_{i-1} = (y^{-1}xy)xZ_{i-1}$, and so $[x^e,y]Z_{i-1} = x^{-e}(y^{-1}xy)^e Z_{i-1} = [x,y]^e Z_{i-1} = Z_{i-1}$, by induction. Thus $[x^e,y] \in Z_{i-1}$ for all $y \in G$. This implies $x^e \in Z_i$ and the induction step is proved. ■

THEOREM 6.5. *Let G be an irreducible linear group of degree n over an algebraically closed field F. If G is nilpotent of class k, then:*

(1) $|G{:}Z(G)|$ *divides* $n^{(k+1)(n-1)}$.

(2) $|G'|$ *divides* $n^{(k+1)(n-1)}$.

Proof. First suppose $\det x = 1$ for all $x \in G$; that is, $G \subseteq SL(V)$ for some vector space V over F. By Corollary 2.6, $Z(G)$ is a group of scalars, so $Z(G) \subseteq \{\zeta 1 \mid \zeta \in F,\ \zeta^n = \det(\zeta 1) = 1\}$; hence $Z(G)$ has exponent e dividing n. By Lemma 6.5 we conclude $x^{e^k} = 1$ for all $x \in G$. Now G is a monomial group by Theorem 4.6B and so G has a normal abelian subgroup A such that G/A is isomorphic to a permutation group of degree n. By Corollary 2.6, A corresponds to a subgroup of $\mathrm{Diag}(n,F)$ over a suitable basis for V. Since A has exponent dividing e^k and each element of A has determinant 1 we conclude that $|A|$ divides $(e^k)^{n-1}$. On the other hand, $|G/A|$ divides $\prod_p p^{\nu_p}$ where $\nu_p = [n/p] + [n/p^2] + \ldots \leqslant n-1$ is the exponent to which the prime p divides $n!$. Therefore $|G| = |G/A| \cdot |A|$ divides $e^{k(n-1)}e^{n-1}$, and hence $|G|$ divides $e^{(k+1)(n-1)}$. Thus proves the theorem in this case.

To complete the proof in the general case we proceed as follows. Let S be the group of all scalars in $GL(V)$, and put $G_0 = GS \cap SL(V)$. Then G_0 is an irreducible linear group which is nilpotent of class k.

Therefore, from the case above, $|G_0|$ divides $n^{(k+1)(n-1)}$. Since $G_0S = GS$, $S \cap G = Z(G)$ and $S \cap G_0 = Z(G_0)$, we conclude that $|G : Z(G)| = |GS : S| = |G_0 : Z(G_0)|$ and (1) is proved. Finally since $\det[x, y] = 1$ for all $x, y \in GL(V)$, $G' \subseteq G \cap SL(V) \subseteq G_0$ and so $|G'|$ divides $|G_0|$ and (2) is proved. ■

COROLLARY 6.5. *Let G be a completely reducible linear group of degree n. If G is nilpotent of class k, then* $|G:Z(G)|$ *and* $|G'|$ *are each at most* $n^{(k+1)(n-1)}$. *Moreover, if G is a subgroup of the special linear group, then* $|G| \leqslant n^{(k+1)(n-1)}$.

Proof. From Corollary 2.8C we may suppose that the underlying field is algebraically closed. The result now follows in an obvious way from the Theorem. ■

Exercises. 1. Let p be a prime and let ζ be a primitive p^mth root of 1 in $\mathbf{C}$ (with $m \geqslant 2$). Let G be the subgroup of $GL(p, \mathbf{C})$ generated by

$$\begin{bmatrix} 0 & 1 & 0 & \dots & 0 \\ 0 & 0 & 1 & \dots & 0 \\ \cdot & \cdot & \cdot & \dots & \cdot \\ \cdot & \cdot & \cdot & \dots & \cdot \\ 0 & 0 & 0 & \dots & 1 \\ 1 & 0 & 0 & \dots & 0 \end{bmatrix} \quad \text{and} \quad \operatorname{diag}(\zeta, 1, 1 \dots, 1).$$

Show that $|G| = p^{(p^m)+1}$ and G is nilpotent of class m. Find $|G:Z(G)|$ and $|G'|$.

2. Show that a completely reducible nilpotent linear group over a field of characteristic $p > 0$ has no non-trivial p-element.

3. Let G be a group with lower central series

$$G = \gamma_1(G) \supseteq \gamma_2(G) \supseteq \dots .$$

If $\gamma_1(G)/\gamma_2(G)$ has exponent e, show that $\gamma_i(G)/\gamma_{i+1}(G)$ has exponent dividing e for $i = 1, 2, \ldots$.

§6.6 In this section we prove a theorem which shows how the 'local' behaviour of a solvable linear group determines whether the group is completely reducible.

Let V be a vector space over an algebraically closed field F. We recall that $x \in GL(V)$ is semisimple if x corresponds to a diagonal matrix over a suitable basis for V.

Note 1. By examining the Jordan canonical form of x we see at once that: x is semisimple $\Longleftrightarrow$ the minimal polynomial for x has no multiple roots.

2. x is semisimple $\Longleftrightarrow$ the cyclic group $\langle x \rangle$ is completely reducible.

LEMMA 6.6. *Let* $x \in GL(V)$ *where* V *is a vector space over an algebraically closed field* F. *Let* h *be an integer* > 0, *and suppose that if* char $F > 0$, *then* char $F \nmid h$. *If* x^h *is semisimple, then so is* x.

Remark. It is clear that if x is semisimple x^h is always semisimple.

Proof. Let $f(X)$, $g(X)$ be the minimal polynomials for x^h and x, respectively. By the Note above, $f(X)$ has no multiple root, and so the greatest common divisor $(f(X), f'(X)) = 1$ where $f'(X)$ is the derived polynomial. Since x^h has no eigenvalue equal to 0, we conclude $(f(X^h), hX^{h-1}f'(X^h)) = 1$. Hence $f(X^h)$ has no multiple root. Since $g(X) \mid f(X^h)$, $g(X)$ has no multiple root and so x is semisimple. ■

The following theorem was proved by Suprunenko (1961) and independently by the author (1963).

THEOREM 6.6. *Let G be a solvable subgroup of $GL(V)$ where V is a vector space over an algebraically closed field F.*

(1) *If each element of G is semisimple, then G is completely reducible.*

(2) *If G is completely reducible, and G has a normal abelian subgroup A of finite index (see Theorem 6.4) such that either char $F = 0$ or char $F \nmid |G:A|$, then each element of G is semisimple.*

Remark. See also Exercise 5 of §8.6.

Proof. (1) By Corollary 6.4 there is a subgroup H of finite index in G such that H' consists of unipotent elements. But a unipotent element is semisimple only if it equals 1. Therefore $H' = 1$ and so H is abelian. Let $A \supseteq H$ be a maximal abelian subgroup of index h in G. We shall now prove that A is completely reducible and that char $F \nmid h$; the result will then follow from Theorem 2.3.

Firstly we show that A is completely reducible; indeed we shall show that V is a direct sum of one-dimensional A-spaces. If A consists of scalers, this is trivial so suppose $x \in A$ is not scalar and proceed by induction on the degree. Let λ be an eigenvalue for x, and define $V_1 = \{v \in V | vx = \lambda v\}$, $V_2 = V(x-\lambda 1)$. Then clearly V_1, V_2 are A-spaces because x centralizes A, and we shall show $V = V_1 \oplus V_2$. Since $V_1 = \ker(x-\lambda 1)$, we have $\dim V - \dim V_1 = \dim V_2$. On the other hand, if $v \in V_1 \cap V_2$, then $v = w(x-\lambda 1)$ for some $w \in V$ and $0 = vx - \lambda v = w(x-\lambda 1)^2$. Because x is semisimple its minimal polynomial has no multiple root and so $w(x-\lambda 1)^2 = 0$ implies $w(x-\lambda 1) = 0$; that is, $v = 0$. Hence $V_1 \cap V_2 = 0$ and we conclude $V = V_1 \oplus V_2$. Since V_1 and V_2 are both proper subspaces of V and both are A-spaces, the induction hypothesis shows that each is a direct sum of one-dimensional A-spaces. This proves the complete reducibility of A.

Secondly, we show that if char $F = p > 0$, then $p \nmid |G:A|$. For otherwise there would exist $y \in G$, $y \notin A$ where $y^p \in A$; to see this, let N be a normal subgroup of G of finite index in A and consider the factor group G/N. Since y is semisimple it corresponds to a matrix $\operatorname{diag}(\eta_1, \ldots, \eta_n)$ over some basis for V. Then y^p corresponds to $\operatorname{diag}(\eta_1^p, \ldots, \eta_n^p)$. But because char $F = p$, $\eta_i^p - \eta_j^p = (\eta_i - \eta_j)^p$ and so $\eta_i^p = \eta_j^p \Longleftrightarrow \eta_i = \eta_j$. Therefore $C_G(y) = C_G(y^p) \supseteq A$. But this implies that the group $\langle A, y\rangle$ generated by A and y is abelian which is contrary to the choice of A as a maximal abelian subgroup. Hence we conclude $p \nmid |G:A|$. The complete reducibility of G now follows from Theorem 2.3.

(2) Since G is completely reducible, A is completely reducible by Theorem 2.2. Since A is abelian and F is algebraically closed, Corollary 2.6 shows that V is a direct sum of one-dimensional A-spaces and so A corresponds to a subgroup of $\operatorname{Diag}(n,F)$ over a suitable basis for V. In particular, each element of A is semisimple. Therefore, under the given hypothesis on $|G:A|$, each element of G is semi-simple by Lemma 6.6. ∎

Applying Theorem 2.5 we get an immediate corollary.

COROLLARY 6.6. *Let G be a solvable linear group. Then G is completely reducible $\Longleftrightarrow$ each finitely generated subgroup of G is completely reducible.* ∎

Exercises. 1. Let F be a field of characteristic $p > 0$ with at least three elements. Let G be the subgroup of $GL(p,F)$ generated by

$$u = \begin{bmatrix} 0 & 1 & 0 & \dots & 0 \\ 0 & 0 & 1 & \dots & 0 \\ . & . & . & \dots & . \\ . & . & . & \dots & . \\ . & . & . & \dots & . \\ 0 & 0 & 0 & \dots & 1 \\ 1 & 0 & 0 & \dots & 0 \end{bmatrix} \quad \text{and} \quad \operatorname{diag}(\xi_1, \dots, \xi_p)$$

for all non-zero $\xi_i \in F$. Show G is solvable and completely reducible but u is not semisimple.

2. Let F be a field of characteristic $p > 0$, and let F_0 be its prime subfield. Let G be the subgroup of $GL(2,F)$ generated by

$$\begin{bmatrix} 1 & 1 \\ 0 & 1 \end{bmatrix} \quad \text{and} \quad \begin{bmatrix} \xi & 0 \\ 0 & \eta \end{bmatrix}$$

for all non-zero $\xi, \eta \in F$ with $\xi\eta^{-1} \in F_0$. Show that G is solvable with a completely reducible subgroup of index p, but G is not completely reducible.

§6.7 *Notes and References*

The first study of solvable linear groups (over **C**) appears in

Zassenhaus, H. Beweis eines Satzes über diskrete Gruppen, *Abh. math. Semin. Univ. Hamburg,* **12**, 289–312 (1938).

Theorem 6.2 appears there with a less precise bound. The paper

Huppert, B. Lineare auflosbare Gruppen, *Math. Z.* **69**, 479–518 (1957)

contains a great deal of information about solvable linear groups. Huppert is interested in applying his results to analyse the structure of finite (abstract) solvable groups and so is particularly interested in the case where the characteristic is not 0. Not only does he prove Theorem 6.2 but he shows that the solvable length ℓ is bounded by a linear function of $\log n$. Recent work of my own (see the reference in §2.11) shows that when G is a completely reducible solvable linear group of degree n, then its solvable length ℓ satisfies $3^{(2\ell-5)/5} \leqslant n$; there is a similar bound when G is not completely reducible.

The exercises of §6.1 are due to Huppert. Exercise 2 of §6.2 is unpublished work of my own. Generally, for each integer $k \geqslant 1$, there is a constant c_k such that if G is a finite solvable completely reducible linear group of degree n, and each prime $p \mid |G|$ satisfies $c_k p^k > n$, then $G^{(2k-1)} = 1$.

Theorem 6.4 appeared in a paper which deals with the structure of (abstract) solvable groups with certain finiteness conditions. See

Mal'cev, A. I. On some classes of infinite solvable groups, *Mat. Sb.* N.S. **28** (70), 567–588 (1951) (= *Amer. Math. Soc. Transl.* (2), **2**, 1–22 (1956)).

The exercise of §6.4 is due to O. H. Kegel.

Over the last 15–20 years, D. A. Suprunenko and several of his students have written numerous papers on linear groups. Some of these results (up to 1958) appear in

Suprunenko, D. A. Soluble and Nilpotent Linear Groups, *Transl. Math. Monographs*, Vol. 9, Amer. Math. Soc., Rhode Island (1963).

The material of §6.5 is from this book.

The material of §6.3 and §6.6 is from the papers

Dixon, J. D. The Fitting subgroup of a linear solvable group, *J. Aust. math. Soc.* **7**, 417–424 (1967).

——, Normal p-subgroups of solvable linear groups, ibid, 545–551.

The first of these papers gives a sharpened form of Theorem 6.4.

Finally we might note a few facts about solvable matrix groups over the ring **Z** and other rings of algebraic integers. It can be shown that such groups are *polycyclic* (that is, they have normal series with cyclic factors), and that the number of infinite cyclic factors in a normal series is bounded by a function of the degree. Some results may be found in

Dade, E. C. Abelian groups of unimodular matrices, *Illinois J. Math.* **3**, 11–27 (1959).

Further results have been proved by Mary R. Freislich (University of New South Wales) in unpublished work. Moreover, there is an appropriate converse to these results: every polycyclic group has a faithful matrix representation over **Z**; see

Auslander, L. On a problem of Philip Hall, *Ann. Math.* **86**, 112–116 (1967).

CHAPTER 7

P-Solvable Groups

§7.1 Let p be a prime. A *p-solvable group* G is a finite group with a normal series

$$G = G_0 \supseteq G_1 \supseteq \ldots \supseteq G_m = 1 \qquad (7.1.1)$$

in which each factor group G_{i-1}/G_i is either a p-group or a p'-group (that is, a group of order prime to p).

Notes 1. Each finite solvable group is p-solvable for each prime p.

2. Since Feit and Thompson have shown that groups of odd order are always solvable, we know that 2-solvable groups must be solvable.

3. G is called *p-solvable of length* ℓ if the series (7.1.1) has $m \leqslant 2\ell$ and no similar series has as few as $2(\ell-1)$ terms. Clearly in a series of this type of minimal length the factors are alternately p-groups and p'-groups.

The following lemma plays an important part in the study of p-solvable groups. We shall not stop to prove it here but refer to M. Hall, *Group Theory,* MacMillan, New York (1959) where the result is proved in Theorems 18.4.4 and 18.4.5.

LEMMA 7.1. *Let G be a p-solvable group.*

(1) *Suppose G has no non-trivial normal p-subgroup, and let N*

be the (unique) largest normal p'-subgroup of G. Then $C_G(N) \subseteq N$.

(2) *Suppose that G has no non-trivial normal p'-subgroup, and le P be the largest normal p-subgroup of G. Then $C_G(P/\Phi(P)) = P$.* ■

Note. Here $\Phi(P)$ is the Frattini subgroup of P and so by the 'Burnside Basis Theorem' $P/\Phi(P)$ is an elementary abelian p-group of order p^d, say. Thus in part (2) of the Lemma we can conclude that G/P has a faithful representation of degree d over the field wit p elements (compare with Theorem 4.4).

The present chapter deals with the structure of p-solvable groups, beginning in §7.2 with a theorem of Huppert. In §7.3 we give a partial generalization of Theorem 6.3 to p-solvable groups. The remainder of the chapter is given over to the proof of our main theorem (Theorem 7.5) widely known as 'the Hall–Higman Theorem B'. This theorem was first proved by P. Hall and G. Higman as a tool in their study of Burnside's problem, and it has since had many applications. The eigenvalues of a p-element of a linear group over a field of characteristic p are all equal to 1, and so give no useful information about the element. The Hall–Higman theorem therefore studies the minimal polynomial of a p-element in a p-solvable linear group over a field of characteristic p and this gives us a surprising amount of information. The proof of this result will use several of our earlier structure theorems.

§7.2 The present section deals with a result of Huppert (1965).

THEOREM 7.2. *Let G be a p-solvable irreducible subgroup of $GL(n, F)$ where F is a finite field of order p^r (and hence* char $F = p$*) If the order g of G is relatively prime to n, then G is cyclic and $g | p^{nr} - 1$.*

Proof. We first use induction on g to prove that G is a cyclic p'-group. Let M be a maximal normal subgroup of G. By Theorem 2.2 the irreducible components of M all have the same degree d which divides n. Thus by induction the irreducible components of M are cyclic p'-groups, and so M is an abelian p'-group.

Now using Corollary 3.4A we can choose E as a finite extension field of F such that each irreducible component of G over E is absolutely irreducible. Then for some $m|n$, G is isomorphic to an absolutely irreducible subgroup G_1 of $GL(m, E)$ by Theorem 2.10B; let M_1 be the subgroup of G_1 corresponding to M. Let $\overline{E}$ be the algebraic closure of E; then G_1 is an irreducible subgroup of $GL(m, \overline{E})$.

Now suppose $p|g$. Because $p \nmid |M|$ and G is p-solvable we have $|G : M| = |G_1 : M_1| = p$. Since M_1 is abelian, G_1 is solvable. Since $(m, g) = 1$, G_1 is primitive by Theorem 4.2B. Then from Theorems 4.5 and 4.4 (3) we conclude $m = 1$. Thus G_1 is a subgroup of the multiplicative group of non-zero elements of $\overline{E}$; hence $g = |G_1|$ is prime to p contrary to our supposition. Thus $p \nmid g$.

Since $p \nmid g$, the degree m of G_1 divides the order g by Theorem 5.2. But $(m, g) = 1$ because $m \mid n$, and so $m = 1$. Hence G_1 is a subgroup of the multiplicative group of non-zero elements of E. The latter is a cyclic p'-group, and so $G(\simeq G_1)$ is also a cyclic p'-group.

Finally we show $g \mid p^{nr} - 1$. By Schur's lemma (Theorem 2.6(2)) the centralizer of G in $M(n, F)$ is a division ring. Since $L = FG$ is a finite commutative subring of this centralizer, L is a field. If x generates G and x has a minimal polynomial of degree d ($\leqslant n$) over F, then $\dim_F L = d$; indeed, $L = \bigoplus_{i=0}^{d-1} Fx^i$. Let V be the underlying vector space for G. Then for each $v \neq 0$ in V, $vL = V$ because G is irreducible. Therefore $d = \dim L \geqslant \dim V = n$; hence $d = n$. Thus the multiplicative group of non-zero elements of L is of order p^{nr} – Since G is a subgroup of this group, we conclude $g \mid p^{nr} - 1$. ■

§7.3 Recall that the *p-core* of a finite group is the (unique) larges normal p-subgroup of that group. Theorem 6.3 showed that the p-core of a solvable linear group is in some sense large. The theorem of this section generalizes this result in part to p-solvable groups.

We shall need two general lemmas.

LEMMA 7.3A. *Let G be a finite group and let H be a normal p'-subgroup whose index is a power of p. Let P be a Sylow p-group of G. Then for each prime q dividing $|H|$, there is a Sylow q-group Q of H such that $P \subseteq N_G(Q)$*

Proof. Let $\Omega = \{Q_1, \ldots, Q_m\}$ be the set of all Sylow q-groups of H; note that $m = |H : N_H(Q_i)|$ and so $m \mid |H|$. Then for each $x \in$ there is a permutation $Q_i \mapsto x^{-1}Q_ix$ of Ω, and in this way P acts as a permutation group on Ω. From elementary theory of permutation groups we know that the orbits on Ω each have order dividing $|P|$, and so each orbit has order a power of p. Since $p \nmid |H|$, $p \nmid m$, an so one orbit, say $\{Q_i\}$, has order 1. Then $Q = Q_i$ has the required property. ■

LEMMA 7.3B. *Let G be a p-solvable group, and let K be the p-core of G. Let H/K be the largest normal p'-subgroup of G/K. If P is a Sylow p-group of G, then K is the p-core of PH.*

Proof. Let L be the p-core of PH; clearly $K \subseteq L$. Since L/K and H/K are normal subgroups of relatively prime order in PH/K, $LH/K = L/K \times H/K$; hence $L/K \subseteq C_{G/K}(H/K)$. But G/K has p-core 1, so $L/K \subseteq H/K$ by Lemma 7.1. Therefore $L = K$. ■

THEOREM 7.3. *Let G be a p-solvable linear group of degree n over a field F of characteristic 0 or p. Then:*

(1) *The Sylow p-group of G is normal if either (a) $p > n+1$, or (b) $p > n$ and p is not a Fermat prime.*

(2) *If the Sylow p-groups of G are cyclic, and K is the p-core of G, then the index of K in a Sylow p-group of G is $\leqslant p^{\lambda_p(n)}$ where $\lambda_p(n)$ is defined in §6.3.*

Proof. (1) Let K be the p-core of G and let P be a Sylow p-group of G; we must prove $P = K$. Let H/K be the largest normal p'-subgroup of G/K. Then PH/K has p-core 1 by Lemma 7.3B and so Lemma 7.3A shows that for each prime $q \mid |H/K|$ there is a Sylow q-group Q of H such that P/K normalizes QK/K; in particular PQ is a solvable linear group of degree n. If char $F = 0$, then PQ is completely reducible by Theorem 2.3, and so $Q \subseteq N_G(P)$ by Theorem 6.3. If char $F = p$, then the same conclusion follows by using Corollary 6.3. Thus in either case we conclude that for each prime q dividing $|H/K|$ there is a Sylow q-group of H which normalizes P. Hence $H \subseteq KN_G(P)$. In particular, P/K is normal in PH/K and so P/K is the p-core of PH/K. But we have seen that the p-core of PH/K is 1. Therefore $P = K$ and P is normal in G.

(2) Using the notation of (1) a similar argument shows that for each prime q dividing $|H/K|$ we have a Sylow q-group Q of H normalized by P; and the solvable group PQ has p-core K_q with $|P : K_q| \leqslant p^{\lambda_q}$ where $\lambda_q \leqslant \lambda_p(n)$ by Theorem 6.3 and its Corollary. Since P is cyclic it has at most one subgroup of any given order. Thus each subgroup of K_q is characteristic in K_q and hence normal in PQ. Hence, if K_0 is the unique subgroup of index $p^{\lambda_p(n)}$ in P, then K_0 is normalized by Q. This is true for each prime q dividing $|H/K|$, and so $H \subseteq KN_G(K_0)$. Hence KK_0/K is in the p-core of PH/K

which is 1 by Lemma 7.3B. Thus $K_0 \subseteq K$ and $|P:K| \leqslant |P:K\ | = p^{\lambda_p(r}$ as required. ■

§7.4 The object of the next two sections is a proof of a theorem due to P. Hall and G. Higman (1956). The present section gives a preliminary lemma and the proof of the Hall–Higman theorem in a special case. The general case is dealt with in §7.5.

The following lemma is due to J. A. Green (1956). It is a useful supplement to Theorem 4.2A.

LEMMA 7.4. *Let G be a finite irreducible subgroup of GL(V) where V is a vector space over a field F of characteristic* $p > 0$. *Let N be a normal subgroup of G where* $|G : N|$ *is a power of p. If all minimal N-spaces of V are N-isomorphic, then N is irreducible.*

Remark. Thus it follows from Theorem 4.2A that if N is not irreducible then G is not primitive.

Proof. By Theorem 3.4B we may assume that F is a finite field. We first note that if W and W' are minimal N-spaces in V, then $\mathrm{Hom}_{FN}(W, W')$ (the set of all N-homomorphisms of W into W') is a vector space of finite dimension over F, and so it has order $p^r > 1$ for some integer r. Since all minimal N-spaces are N-isomorphic, r is independent of the choice for W and W'.

Now $V = \bigoplus_{i=1}^{s} V_i$ where the V_i are minimal N-spaces by Theorem 2.2; we have to prove $s = 1$. We shall begin by counting the number n_t of minimal N-spaces contained in $\bigoplus_{i=1}^{t} V_i$ for $t = 1, \ldots, s$; clearly $n_1 = 1$. If $t > 1$, then a minimal N-space W contained in $\bigoplus_{i=1}^{t} V_i$ either equals V_t or else is of the form $\{u + u\alpha \mid u \in W\}$ where W is a minimal N-space in $\bigoplus_{i=1}^{t-1} V_i$ and $\alpha \in \mathrm{Hom}_{FN}(W, V_t)$. Since W can be chosen in

n_{t-1} ways and α can be chosen in p^r ways, therefore $n_t = 1 + p^r n_{t-1}$. Hence we have $n_t = 1 + p^r + \ldots + p^{r(t-1)}$ for $t = 1, \ldots, s$.

Now consider the action of G in permuting the n_s minimal N-spaces in V. Each of these N-spaces is fixed by N, so by elementary theory of permutation groups the orbits of G have order dividing $|G : N|$. Thus each of these orbits is a power of p. Since $n_s \equiv 1 \pmod p$ at least one orbit, say $\{W\}$, has length 1. Then W is a G-space, so $W = V$ because G is irreducible. Hence V is a minimal N-space and N is irreducible as required. ■

THEOREM 7.4. *Let G be a finite primitive linear group over an algebraically closed field F of characteristic $p > 0$. Suppose that:*

(1) *$G = PQ$ where $P = \langle x \rangle$ is a cyclic group of order p^m, and Q is a normal Sylow q-group of G with $q \neq p$.*

(2) *Q is an irreducible group, and whenever Q_0 is a proper subgroup of Q such that $x^{-1}Q_0x = Q_0$ we have $C_{Q_0}(x^{p^{m-1}}) = Q_0$.*

Then G is of degree q^n for some integer $n \geqslant 0$, and the minimal polynomial for x is $(X-1)^s$ where $s = p^m - 1$ if $p^m = q^n + 1$, and $s = p^m$ otherwise.

Proof. Since F has characteristic p, no non-trivial p-subgroup of G is completely reducible. Thus by Theorem 2.2 the p-core of G is 1, and so $Q = \mathrm{Fit}(G)$. Put $Z = Z(Q)$. By Theorem 4.4, Q/Z is an elementary abelian q-group and $|Q/Z|$ is the square of the degree of G. Thus $|Q/Z| = q^{2n}$ and G has degree q^n for some integer $n \geqslant 0$.

Without changing our notation we shall now go over to the matrix form for G choosing our basis so that x is in Jordan canonical form:

$$x = \mathrm{diag}(x_1, \ldots, x_r) \qquad (7.4.1)$$

where

$$x_i = \begin{bmatrix} 1 & 0 & \dots & 0 \\ 1 & 1 & \dots & 0 \\ \vdots & 1 & \ddots & \vdots \\ 0 & \dots & \dots & 1 \end{bmatrix}$$

is a $d_i \times d_i$ Jordan block ($i = 1, \dots, s$) with $d_1 \geqslant d_2 \geqslant \dots \geqslant d_r$ and $d_1 + \dots + d_r = q^n$. (Note that the eigenvalues of x are all 1 because x is a p-element and char $F = p$.) Because x has order p^m, $d_1 \leqslant p^m$. We shall now calculate the dimension c of the centralizer of x in $M(q^n, F)$ in two ways.

Firstly, using a block decomposition similar to (7.4.1) we write $y \in M(q^n, F)$ in the form

$$y = \begin{bmatrix} y_{11} & y_{12} & \cdots & y_{1r} \\ y_{21} & y_{22} & \cdots & y_{2r} \\ \vdots & & & \\ y_{r1} & y_{r2} & \cdots & y_{rr} \end{bmatrix}.$$

Then $xy = yx \Longleftrightarrow x_i y_{ij} = y_{ij} x_j$ for all $i, j = 1, \dots, r$. Now a $d_i \times d_j$ matrix $z = [\zeta_{k\ell}]$ satisfies $x_i z = z x_j$ exactly when:

$$\zeta_{k\ell} = \zeta_{k+1,\ell+1} \quad \text{for } \ell \leqslant k \leqslant \min(d_i, d_j) - 1$$

and

$$\zeta_{k\ell} = 0 \quad \text{for } \ell > k.$$

Thus there are $\min(d_i, d_j)$ linearly independent solutions z to $x_i z = z x_j$. Hence

$$c = \sum_{i=1}^{r} \sum_{j=1}^{r} \min(d_i, d_j) = \sum_{i=1}^{r} (2i-1)d_i \qquad (7.4.2)$$

since $d_1 \geqslant \ldots \geqslant d_r$.

Secondly, because Q is irreducible, $FQ = M(q^n, F)$ by Theorem 2.6B. Thus, if $y_1, \ldots, y_{q^{2n}}$ is a set of coset representatives for Z in Q, then these elements span FQ because Z is a group of scalars, and so they form a basis for FQ over F. Now an element $\sum_i a_i y_i$ $(a_i \in F)$ centralizes $x \Longleftrightarrow a_i y_i = a_j x^{-1} y_j x$ whenever $y_i Z = x^{-1} y_j x Z$. Thus the dimension c of the centralizer of x equals the number of classes of cosets in Q/Z conjugate under powers of x. We shall now show that with the exception of $\{Z\}$, each of these classes contains p^m cosets and it then follows that

$$c = 1 + (q^{2n} - 1)/p^m. \qquad (7.4.3)$$

Consider $C/Z = C_{Q/Z}(x^{p^{m-1}})$. Since $C \supseteq Z \supseteq Q'$, C is normalized by both Q and x, and so C is normal in G. Now P acts by conjugation on the elementary abelian q-group Q/Z; this corresponds to the action of a linear p-group $\overline{P}$ on a vector space $\overline{Q}$ over a field with q elements. Thus Theorem 2.3 shows that the $\overline{P}$-space corresponding to C in $\overline{Q}$ has a complementary $\overline{P}$-space in $\overline{Q}$; that is, there exists a subgroup Q_0/Z of Q/Z such that $Q = Q_0 C$ and $Q_0 \cap C = Z$ with $x^{-1} Q_0 x = Q_0$. Since $C_{Q_0}(x^{p^{m-1}}) \subseteq Q_0 \cap C = Z$, the hypothesis (2) on Q shows that $Q_0 = Q$ or $Q_0 = Z$. If the latter held, C would equal Q and then $x^{p^{m-1}}$ centralizes Q/Z contrary to Theorem 4.5 (6) because $Q = \mathrm{Fit}(G)$. Thus $Q_0 = Q$ and $C = Z$. Thus no coset in Q/Z different from Z is centralized by $x^{p^{m-1}}$ and this shows that each coset $\neq Z$ in Q/Z has p^m conjugates under P. Hence (7.4.3) follows.

It now follows from (7.4.2) and (7.4.3) that

$$\sum_{i=1}^{r} (2i-1)d_i = 1 + (q^{2n}-1)/p^m \qquad (7.4.4)$$

where $p^m \geqslant d_1 \geqslant \ldots \geqslant d_r > 0$ and $d_1 + \ldots + d_r = q^n$. Write $q^n = kp^m + h$ where $0 < h < p^m$. Then the right-hand side of (7.4.4) equals

$$1 + \{(kp^m + h)^2 - 1\}/p^m = k^2p^m + 2kh + 1 + (h^2-1)/p^m,$$

whilst the left-hand side is at least as large as

$$\sum_{i=1}^{k} (2i-1)p^m + (2k+1)h = k^2p^m + (2k+1)h.$$

Hence we have

$$k^2p^m + (2k+1)h \leqslant k^2p^m + 2kh + 1 + (h^2-1)/p^m,$$

and so $h - 1 \leqslant (h^2-1)/p^m$. But the definition of h, $(h^2-1)/p^m \leqslant h-1$, and so we conclude that we have equality at each step. In particular, this shows that $(h^2-1)/p^m = h-1$ and so $h = p^m - 1$; and that $d_1 = \ldots = d_k = p^m$, $d_{k+1} = p^m - 1$ and so $r = k+1$, because these are the only values of the d_i for which the left-hand side of (7.4.4) attain its minimum value. Finally the minimal polynomial for x is $(X-1)^s$ where $s = d_1$. Therefore $s = p^m$ when $k \geqslant 1$, and $s = p^m - 1 = q^n$ when $k = 0$. ∎

Note. In all cases we have $q^n \equiv -1 \pmod{p^m}$.

§7.5 We now deal with the general case of the Hall–Higman Theorem B.

THEOREM 7.5. *Let G be a p-solvable linear group over a field F of characteristic p, and suppose that G has p-core 1. Let $x \in G$*

have order $p^m \geqslant 1$. *Then the minimal polynomial for* x *is* $(X-1)^s$ *where:*

(1) $s = p^m$ *if* p *is odd and not a Fermat prime.*

(2) $s \geqslant p^{m-1}(p-1)$ *if* p *is a Fermat prime; moreover* $s = p^m$ *if the Sylow 2-groups of* G *are abelian.*

(3) $s \geqslant 2^{m-m_0}(2^{m_0}-1) \geqslant 2^{m-2}\,3$ *if* $p = 2$ *and* $2^{m_0}-1$ *is the largest Mersenne prime for which* G *has a non-abelian Sylow subgroup; moreover* $s = 2^m$ *if each Sylow subgroup corresponding to a Mersenne prime is abelian.*

Remark. A *Mersenne prime* is a prime of the form $2^k - 1$. We shall call x *exceptional* when $s \neq p^m$. Note that $s \leqslant p^m$ because $(x-1)^{p^m} = x^{p^m} - 1 = 0$ since char $F = p$.

Proof. We proceed by induction on (i) the degree and (ii) the order of G.

(A) Since G is p-solvable with p-core equal to 1, the largest normal p'-subgroup N of G is not 1. The p-core K of $N\langle x\rangle$ must centralize N because $[N, K] \subseteq N \cap K = 1$. Thus $K = 1$ by Lemma 7.1, and so by induction we may suppose $G = N\langle x\rangle$. Note that, if $m \geqslant 1$, then $x^{p^{m-1}}$ does not centralize N (by Lemma 7.1).

(B) Clearly we may suppose F is algebraically closed. Moreover we can suppose G is completely reducible because the process described in Theorem 2.4 cannot increase the degree of the minimal polynomial. We may even suppose that G is irreducible. Indeed, if the underlying space $V = V_1 \oplus V_2$ where V_1, V_2 are non-trivial G-spaces, then at least one of $x|V_1$, $x|V_2$ is of order p^m; suppose $x|V_1$ has order p^m. Since $G|V_1$ is completely reducible, it has p-core 1 because char $F = p$, and so by the induction hypothesis $x|V_1$ has a minimal polynomial $(X-1)^{s'}$ where s' satisfies the conditions (1)–(3) in place of s. Since $s \geqslant s'$, s also satisfies

(1)–(3). Thus the only case we need consider is when G is irreducible over an algebraically closed field.

(C) We may suppose that N (see (A)) is a q-group for some prime $q \neq p$. Indeed, let $q_1, \ldots, q_t$ be the primes which divide $|N|$. Then from Lemma 7.3A, for each i there is a Sylow q_i-group Q_i of N which is normalized by $\langle x \rangle$. Let K_i be the p-core of $Q_i \langle x \rangle$ and note that $[Q_i, K_i] \subseteq Q_i \cap K_i = 1$ and hence K_i centralizes Q_i. If $K_i \neq 1$, then $x^{p^{m-1}} \in K_i$. Since $x^{p^{m-1}}$ does not centralize N by (A), there must be some Q_j which is not centralized by $x^{p^{m-1}}$. Then $Q_j \langle x \rangle$ has p-core $K_j = 1$. Now choose $Q \subseteq Q_j$ minimal with respect to the condition that $Q\langle x \rangle$ has p-core 1. Then $Q\langle x \rangle$ satisfies the conditions of the theorem and so by induction we may suppose $G = Q\langle x \rangle$; in particular, G is solvable.

(D) Now consider the case where G is imprimitive. Then the underlying vector space $V = \bigoplus_{i=1}^{\ell} W_i$ ($\ell \geqslant 2$) where the set of subspace W_i is permuted under the action of G. Let H be the (normal) subgroup of G consisting of all $y \in G$ such that $W_i y = W_i$ for all i. Let x^{p^t} be the smallest power of x in H. By re-ordering the W_i if necessary, we may suppose $W_1 x^j \neq W_1$ for $j = 1, \ldots, p^t - 1$. Now $H|W_1$ is completely reducible by Theorem 2.2 so its p-core is 1. Thus, by induction, the minimal polynomial for $x^{p^t}|W_1$ has degree s' where s' satisfies the conditions (1)–(3) with $m' = m - t$ in place of m. Write the degree s of the minimal polynomial for x in the form $s = hp^t + k$ $(0 \leqslant k < p^t)$. Then for any $v \in W_1$, $v(x-1)^{(hp^t)+k} = 0$. and so

$$0 = v(x^{p^t} - 1)^h (x-1)^k = \sum_{i=0}^{k} (-1)^{k-i} \binom{k}{i} v_1 x^i \qquad (7.5.1)$$

where $v_1 \overset{\text{def}}{=} v(x^{p^t} - 1)^h \in W_1$ because $x^{p^t} \in H$. Since $W_1 x^i = W_{j_i}$ for some $j_i \neq 1$ for $i = 1, \ldots, p^t - 1$, we conclude that the first term in the sum in (7.5.1) is linearly independent of the other terms; hence $v_1 = 0$. Thus for each $v \in W_1$, $v(x^{p^t} - 1)^h = 0$, so the degree s' of the minimal polynomial for $x^{p^t}|W_1$ is $\leqslant h$. Therefore $s = hp^t + k \geqslant s'p^t$. Because s' satisfies the conditions (1)–(3) with $m' = m-t$ in place of m, we conclude that s satisfies these conditions in terms of m.

(E) Finally, we consider the case where G is primitive. In this case Q is irreducible by Lemma 7.4. From the choice of Q in (C) we see that the hypotheses of Theorem 7.4 are satisfied. The theorem thus follows in this case when we observe that in the exceptional cases Q must be a non-abelian group. ■

Exercise. Show that the theorem remains valid (in the sense that s is the degree of the minimal polynomial of x although this polynomial is not of the form $(X-1)^s$) when F has characteristic 0. [*Hint:* Use §3.6.]

§7.6 *Notes and References*

For a general reference to the subject of p-solvable groups see the reference to M. Hall's book in §7.1 or p. 688–696 of Huppert's book.

The theorem of §7.2 appears in

Huppert, B. Zur Gaschützschen Theorie der Formationen, *Math. Annln.* **164**, 133–141 (1966).

This reference also gives a number of applications of the theory of linear groups.

Theorem 7.3 is a modest generalization of a result of Itô; see the reference quoted in §6.7. Part (2) of this theorem was proved independently (for the field **C**) by D. L. Winter. He has also published

Winter, D. L. On finite linear groups, *Math. Z.* **106**, 245-246 (1968)

which includes some related results. See also the paper of Glauberman referred to in §3.9.

Theorem 7.5 appears as Theorem B in

Hall, P., and Higman, G On the *p*-length of *p*-soluble groups and reduction theorems for Burnside's problem, *Proc. Lond. math. Soc.* **7**, 1–42 (1956).

Lemma 7.4 and Theorem 7.4 also appear there. Our proofs follow the pattern of the original proofs but we should note that there have been several shorter proofs due to Thompson, to Feit, and to others. We refer to the Hall–Higman paper for some applications of Theorem 7.5.

CHAPTER 8

Zariski Topology and Algebraic Groups

§8.1 Although we shall not be looking at topological groups we shall introduce in this chapter a special kind of topology (the Zariski topology) onto an arbitrary matrix group. This is a comparatively weak topology, for example it is not usually Hausdorff, but it does have certain convenient properties. A subgroup of the general linear group which is a closed subgroup under the Zariski topology is called an algebraic group; for example most of the special groups introduced in Chapter 1 are algebraic groups. Semitopological methods prove very powerful in analyzing algebraic groups, and moreover they often permit us to give unified proofs for theorems which classically required consideration of many special cases. What is given here is at best a superficial introduction to a widely developed subject, and we refer to the references quoted in §8.9 for further reading.

In §8.2–§8.4 we introduce the basic concepts and elementary properties of the Zariski topology. Beginning with §8.5 we examine the relation of this topology to the group structure of a matrix group. Theorem 8.6 completely characterizes the connected solvable linear groups over an algebraically closed field. Finally in §8.7 and §8.8 we show that an algebraic group over an algebraically closed field contains the unipotent and semisimple parts of each of its elements.

Note. For this chapter, F will always denote an algebraically closed field (of arbitrary characteristic).

§8.2 Let F be an algebraically closed field. We write F^n to denote the vector space over F of all n-tuples $x = (\xi_1, \ldots, \xi_n)$ with $\xi_i \in F$. Let $X = (X_1, \ldots, X_n)$ be a family of n indeterminates and consider the ring $F[X] = F[X_1, \ldots, X_n]$ of polynomials over F in these indeterminates. If $f(X) = f(X_1, \ldots, X_n) \in F[X]$ then we write $f(x) = f(\xi_1, \ldots, \xi_n)$.

DEFINITION. A set $A \subseteq F^n$ is *algebraic* if there is a family Δ of polynomials in $F[X]$ such that A is the set of all common zeros for the polynomials in Δ; that is, $x \in A \Longleftrightarrow f(x) = 0$ for all $f(X) \in \Delta$.

For any set $B \subseteq F^n$, $I(B) = \{f(X) \in F[X] \mid f(b) = 0 \text{ for all } b \in B\}$. Note that $I(B)$ is an ideal in the ring $F[X]$, and if B is an algebraic set, then B is the algebraic set corresponding to $\Delta = I(B)$.

Examples. 1. $n = 2$, $A = \{(\xi, 0), (0, \eta) \mid \xi, \eta \in F\}$ is an algebraic set corresponding to $\Delta = \{X_1^2 X_2\}$. Note that $I(A) = (X_1 X_2)$ is not generated by Δ.

2. The sets ϕ, F^n and $\{x\}$ are algebraic for any $x \in F^n$. $I(\phi) = F[X]$, $I(F^n) = (0)$ and $I(\{x\}) = (X_1 - \xi_1, \ldots, X_n - \xi_n)$.

Note. The Hilbert Basis Theorem (see §0.2) asserts that each ideal in $F[X]$ can be generated by a finite number of polynomials. In particular, each algebraic set can be defined by a finite family Δ.

LEMMA 8.2A. *Let $A_\lambda (\lambda \in \Lambda)$ be a family of algebraic sets in F^n. Then $A_\lambda \cup A_\mu$ and $\bigcap_{\lambda \in \Lambda} A_\lambda$ are algebraic sets. In particular, every finite subset of F^n is algebraic (by Example 2 above).*

Proof. $A_\lambda \cup A_\mu$ is the set of common zeros of the family

$I(A_\lambda)I(A_\mu)$; and $\bigcap_{\lambda \in \Lambda} A_\lambda$ is the set of common zeros of the family $\bigcup_{\lambda \in \Lambda} I(A_\lambda)$. ■

LEMMA 8.2B. *If $A_1 \supseteq A_2 \supseteq \ldots$ is a descending sequence of algebraic sets in F^n, then for some integer $m \geqslant 1$: $A_m = A_{m+i}$ for all $i \geqslant 0$. (Descending chain condition.)*

Proof. From the hypothesis $I(A_1) \subseteq I(A_2) \subseteq \ldots$. Let $I = \bigcup_{i=1}^{\infty} I(A_i)$. It is readily verified that I is an ideal in $F[X]$, and so by the Hilbert Basis Theorem (see above) it has a finite number of generators, say $f_1(X), \ldots, f_k(X)$. Choose m_j such that $f_j(X) \in I(A_{m_j})$ and put $m = \max\{m_1, \ldots, m_k\}$. Then $I(A_m) \supseteq (f_1(X), \ldots, f_k(X)) = I$. Hence $I(A_m) = I(A_{m+i})$ for all $i \geqslant 0$. ■

Exercise. Show that the descending chain condition (Lemma 8.2B) is equivalent to the minimal condition:

For any non-empty family Σ of algebraic sets in F^n there is an element $M \in \Sigma$ such that no other element of Σ is a subset of M.

§8.3 We now define the *Zariski topology* $\mathcal{Z}$ on the set F^n as follows. A subset $S \subseteq F^n$ is in $\mathcal{Z}$ (that is, S is an *open* set) $\Longleftrightarrow$ the complement $F^n \backslash S$ is an algebraic set. Alternatively, the *closed* sets are precisely the algebraic sets. It follows from §8.2 that $\mathcal{Z}$ has the required properties of a topology: F^n and ϕ are open sets (Example 2), the intersection of two open sets is open, and the union of an arbitrary family of open sets is open (Lemma 8.2A). Moreover $\mathcal{Z}$ is a T_1-topology since singletons $\{x\}$ are closed by Example 2 of §8.2; however, $\mathcal{Z}$ is not Hausdorff (see Exercise 3 below).

In future all topological terms will refer to the Zariski topology on F^n or to the topology induced on the subsets of F^n.

THEOREM 8.3. *Let $S \subseteq F^n$. Then each family of disjoint non-empty subsets which are open in S is finite. It follows that S is a union of a finite number of disjoint connected components.*

Proof. Suppose $(S_i)_i$ is an infinite sequence of non-empty subsets of S which are open in the induced topology of S, and suppose $S_i \cap S_j = \phi$ for all $i, j, i \neq j$. Let A_j be the closure of $B_j = S \setminus \bigcup_{i=1}^{j} S_i$ in F^n. Since B_j is closed in S, $A_j \cap S = B_j$. Since $A_1 \supseteq A_2 \supseteq \ldots$ is a descending sequence of closed sets in F^n, Lemma 8.2B shows that for some m, $A_m = A_{m+1}$ for all $i \geqslant 0$. This implies $B_m = S \cap A_m = S \cap A_{m+1} = B_{m+1}$ and so $S_{m+1} = \phi$ contrary to hypothesis. This proves the first part of the theorem.

The proof of the second part is as follows. Let $x \in S$ and define Σ as the family of all $T \subseteq S$ with $x \in T$ such that T is both open and closed in S; write $\overline{T}$ for the closure of T in F^n. From the Exercise of §8.2 we may choose $S_x \in \Sigma$ such that $\overline{S}_x$, and hence $S_x = S \cap \overline{S}_x$, is minimal. Clearly S_x is unique and is the connected component of S containing x. Thus the connected components of S are disjoint sets open in S, so the result now follows from the first part of the theorem. ■

Exercises. 1. Consider the case $n = 1$, $F^n = F$. Show that the closed sets are precisely the finite subsets of F together with F itself. What are the connected subsets of F?

2. If $F = \mathbf{C}$, show that any subset S of $\mathbf{C}$ is open in the usual topology if it is open in the Zariski topology; but not conversely.

3. Show that any infinite Hausdorff topological space has an infinite family of disjoint open sets; hence no infinite subset S of F^n is a Hausdorff space. [*Hint:* Each infinite Hausdorff space has two disjoint non-empty open sets, at least one of which is infinite (since distinct points have disjoint neighborhoods).]

§8.4 This section is devoted to showing how we may construct functions which are continuous in the Zariski topology.

Let $S \subseteq F^m$ and let $f : S \to F^n$. Then f is a *rational function* if it has the form

$$f(x) = (f_1(x), \ldots, f_n(x)) \quad (\text{where } x = (\xi_1, \ldots, \xi_m) \in S)$$

where each component $f_i(X_1, \ldots, X_m)$ is a rational function; that is, a quotient $p_i(X_1, \ldots, X_m)/q_i(X_1, \ldots, X_m)$ of relatively prime polynomials over F where $q_i(x) \neq 0$ for each $x \in S$. We call the $p_i(X_1, \ldots, X_m)$ the *numerators* of the rational function.

LEMMA 8.4. *Let $\phi \neq S \subseteq F^m$ and let $f : S \to F^n$ be a rational function. Then f is continuous.*

Proof. We have to show that if $A \subset F^n$ is a closed set, then the inverse image $f^{-1}(A)$ is closed in S. Now

$$f^{-1}(A) = \cap \{x \in S \mid gf(x) = 0\}$$

where the intersection is over all $g(X_1, \ldots, X_n) \in I(A)$. But $\{x \in S \mid gf(x) = 0\}$ is the set of all common zeros in S of the polynomials which are numerators of the rational function gf; hence this set is closed in S. Hence $f^{-1}(A)$ is closed in S. ∎

We can identify the set $M(n, F)$ with the vector space F^{n^2} by the mapping $[\xi_{ij}] \mapsto (\xi_{11}\ \xi_{12} \cdots \xi_{nn})$ and so define the Zariski topology on $M(n, F)$. In this chapter, references to topological concepts with respect to $M(n, F)$, or subsets of $M(n, F)$, always refer to the Zariski topology, or the topology that this induces on subsets. The following theorem now follows from Lemma 8.4 and elementary properties of matrices.

THEOREM 8.4. *All the following mappings of $GL(n, F)$ into itself are continuous (for any given $u \in GL(n, F)$).*

(1) $x \mapsto x^{-1}$ (2) $x \mapsto xu$

(3) $x \mapsto ux$ (4) $x \mapsto x^{-1}ux$

In particular, (1)–(3) are homomorphisms of $GL(n, F)$ onto itself. Hence, if S is a subset which is open [respectively: closed, or connected] in $GL(n, F)$, then the sets $S^{-1} = \{x^{-1} | x \in S\}$, Su and uS are also open [respectively: closed, or connected]. ■

DEFINITION. A subgroup G of $GL(n, F)$ which is closed in $GL(n, F)$ is called an *algebraic* (linear) *group*.

Note. 1. Every finite subgroup of $GL(n, F)$ is algebraic (Lemma 8.2A).

2. If G is an algebraic group in $GL(n, F)$ then by Theorem 8.4 $u^{-1}Gu$ is also an algebraic group. Thus if V is a vector space of dimension n over F we can define a subgroup of $GL(V)$ to be an algebraic group if it corresponds to an algebraic group in $GL(n, F)$ over any basis of V.

Exercises. 1. Let G be a group which has defined on it a T_1-topology such that the mapping $(x, y) \mapsto xy$ is a continuous mapping from $G \times G$ (with the product topology) into G. Show that the topology is Hausdorff. In particular, in view of Exercise 3 of §8.3, the mapping $(x, y) \mapsto xy$ for infinite subgroups G of $GL(n, F)$ is *not* continuous with respect to the Zariski topology. [For example, see Bourbaki, *Topology*, Vol. 1, Chapter III, §1.2.]

2. Show that $SL(n, F)$, $Sp(n, F)$, $\mathrm{Diag}(n, F)$, $TL(n, F)$ and $STL(n, F)$ are all algebraic groups by writing down specific sets of polynomials for which these groups are the common zeros in $GL(n, F)$.

3. Let G be a subgroup of $GL(n, F)$ and let S be a subset of G whose closure in G is G. Show that $FG = FS$.

§8.5 This section deals with the elementary properties of matrix groups with respect to the Zariski topology.

LEMMA 8.5. *Let G be a subgroup of $GL(n, F)$ and let H be a subgroup of G.*

(1) *If there is a subset $S \neq \phi$ which is open in G such that $S \subseteq H$, then H is open in G.*

(2) *If H is open in G, then H is closed in G.*

(3) *Let $\overline{H}$ denote the closure of H in $GL(n, F)$. Then the closure $\overline{H} \cap G$ of H in G is a subgroup, and $\overline{H} \cap G$ is normal in G if H is normal in G.*

(4) *If H is closed in G and $|G:H| < \infty$, then H is open in G.*

(5) *If $S \neq \phi$ is a subset closed in G, then $N_G(S)$ is closed in G.*

(6) *For any subset $T \neq \phi$ in G, $C_G(T)$ is closed in G.*

Proof. (1) Let $x \in S$. For each $y \in H$, $Sx^{-1}y$ is open in G by Theorem 8.4 and $y \in Sx^{-1}y \subseteq H$. Therefore H is open in G.

(2) The complement $G \backslash H$ is a union of left cosets of H in G, thus $G \backslash H$ is a union of open sets (Theorem 8.4) and H is closed in G.

(3) To show that $\overline{H} \cap G$ is a subgroup it is enough to show that $\overline{H}$ is. We use Theorem 8.4 several times. First $\overline{H}x$ is closed for all $x \in H$, so $\overline{H} = \overline{Hx} \subseteq \overline{H}x$. Therefore $\overline{H}x^{-1} \subseteq \overline{H}$ for all $x \in H$ and so $\overline{H}H \subseteq \overline{H}$. Hence, for each $y \in \overline{H}$, $yH \subseteq \overline{H}$ and so $y\overline{H} = \overline{yH} \subseteq \overline{H}$. Thus $\overline{H}\,\overline{H} \subseteq \overline{H}$. Similarly $\overline{H}^{-1} \subseteq \overline{H}$, and so $\overline{H}$ is a subgroup. Finally, if u normalizes H, then $\overline{u^{-1}Hu} = u^{-1}\overline{H}u$ and so u normalizes $\overline{H}$. Hence, if H is normal in G, then $\overline{H} \cap G$ is normal in G.

(4) $G \backslash H$ is a union of a finite number of left cosets of H, and so by Theorem 8.4 it is a union of finite number of closed sets. Hence $G \backslash H$ is closed and H is open in G.

(5) $N_G(S)$ is the intersection of the sets $\{x \in G \mid x^{-1}ux \in S\}$ $(u \in S)$ which are closed in G by Theorem 8.4.

(6) For each $u \in T$, $N_G(\{u\})$ is closed by (5). Hence $C_G(T) = \bigcap_{u \in T} N_G(\{u\})$ is closed. ■

The *connected component of* 1 in a matrix group G is the connected component of G containing the element 1.

THEOREM 8.5. *Let G be a subgroup of $GL(n, F)$, and let G_0 be the connected component of 1 in G. Then G_0 is a normal subgroup of finite index in G and the connected components of G are the cosets of G_0 in G. Moreover, if H is a subgroup of finite index which is closed in G, then $G_0 \subseteq H$.*

Proof. For all $x \in G$, $y \in G_0$ both $x^{-1}G_0x$ and $y^{-1}G_0$ are connected by Theorem 8.4; since each of these sets contains 1, $x^{-1}G_0x \subseteq G_0$ and $y^{-1}G_0 \subseteq G_0$. Hence G_0 is normal and a subgroup of G. Again Theorem 8.4 shows that the coset $xG_0 = G_0x$ is the connected component of x for each $x \in G$. Then Theorem 8.3 shows that G_0 has only a finite number of cosets, so $|G : G_0| < \infty$.

Finally, if H is a closed subgroup of finite index in G, then H is open by Lemma 8.5 (4). Thus H is both open and closed and $1 \in H$, so $G_0 \subseteq H$.

Exercises. 1. Let G be a subgroup of $GL(n, F)$. Show that if $|G : G_0| = k$, then each finite conjugacy class in G contains $\leqslant k$ elements. In particular, if G is connected, then each $u \in G$, $u \notin Z(G)$ has infinitely many conjugates in G. [*Hint:* The continuous mapping $x \mapsto x^{-1}ux$ of G cannot increase the number of connected components.]

2. Let $A_1 \subseteq A_2 \subseteq \ldots$ be an ascending sequence of sets in a subgroup G of $GL(n, F)$. Show that for some integer m, $C_G(A_m) = C_G(A_{m+i})$ for all $i > 0$.

3. Let $S_{\mathbf{N}}^*$ denote the restricted symmetric group consisting of all permutations of $\mathbf{N} = \{1, 2, 3, \ldots\}$ which leave all but a finite number of symbols fixed. Use Exercise 2 to show that $S_{\mathbf{N}}^*$ has no faithful representation of any degree over any field.

4. Show that the index $|G : G_0|$ for a subgroup of $GL(n, F)$ is not bounded by a function of n. [*Hint:* Look at the finite subgroups of $GL(n, F)$.]

§8.6 This section deals with how the solvable and nilpotent structure of a matrix group interacts with the Zariski topology.

LEMMA 8.6A. *Let A, B and C be subgroups of $GL(n, F)$ and denote their closures in $GL(n, F)$ by $\overline{A}$, $\overline{B}$ and $\overline{C}$. If the commutator group $[A, B] \subseteq C$, then $[\overline{A}, \overline{B}] \subseteq \overline{C}$.*

Proof, Let $y \in B$. Then $x \mapsto x^{-1}y^{-1}xy$ is a continuous mapping of $GL(n, F)$ into itself by Theorem 8.4; hence the inverse image of the closed set $\overline{C}$ is closed. Since A is contained in this inverse image, $[\overline{A}, y] \subseteq \overline{C}$ for all $y \in B$; hence $[\overline{A}, B] \subseteq \overline{C}$. If we replace B by $\overline{A}$ and A by B and repeat this argument we then find $[\overline{A}, \overline{B}] \subseteq \overline{C}$. ■

LEMMA 8.6B. *Let G be a subgroup of $GL(n, F)$. Then:*

(1) *If G is solvable of length ℓ with derived series*

$$G = G^{(0)} \supseteq G^{(1)} \supseteq \ldots \supseteq G^{(\ell)} = 1$$

then $\overline{G} = \overline{G}^{(0)} \supseteq \overline{G}^{(1)} \supseteq \ldots \supseteq \overline{G}^{(\ell)} = 1$ is a normal series with abelian factors for $\overline{G}$. Hence $\overline{G}$ is solvable of length ℓ.

(2) *If G is nilpotent of class k with lower central series*

$$G = \gamma_1(G) \supseteq \gamma_2(G) \supseteq \ldots \supseteq \gamma_{k+1}(G) = 1$$

then

$$\overline{G} = \overline{\gamma_1(G)} \supseteq \overline{\gamma_2(G)} \supseteq \ldots \supseteq \overline{\gamma_{k+1}(G)} = 1$$

is a normal series with $[\overline{\gamma_i(G)}, \overline{G}] \subseteq \overline{\gamma_{i+1}(G)}$ *for all i. Hence* $\overline{G}$ *is nilpotent of class k.*

(3) *If G is abelian, then* $\overline{G}$ *is abelian.*

Proof. Follows immediately from Lemma 8.5(3) and Lemma 8.6A. ■

We can now characterize the solvable connected subgroups of $GL(n, F)$ in a theorem due to E. R. Kolchin (1948). The result was proved earlier for the case $F = \mathbf{C}$ by S. Lie.

THEOREM 8.6. *Suppose that G is a solvable connected subgroup of* $GL(n, F)$. *Then G is conjugate in* $GL(n, F)$ *to a subgroup of* $TL(n, F)$.

Remark. Recall our blanket assumption that F is algebraically closed. Observe that since $TL(n, F)$ is solvable our theorem implies that, if G is a connected subgroup of $GL(n, F)$, then G is solvable $\Longleftrightarrow$ G is conjugate to a subgroup of $TL(n, F)$.

Proof. From Corollary 6.4 we know that G has a subgroup H of finite index such that $u^{-1}Hu \subseteq TL(n, F)$ for some $u \in GL(n, F)$. Since $TL(n, F)$ is a closed subgroup of $GL(n, F)$ (Exercise 2 of §8.4), $u^{-1}\overline{H}u = \overline{u^{-1}Hu} \subseteq TL(n, F)$. Thus we may suppose H is closed in G by Lemma 8.5(3). Then $G_0 \subseteq H$ by Theorem 8.5. Since $G = G_0$, $G = H$ and the result follows. ■

Exercises. 1. If G is a connected subgroup of $GL(n, F)$ and S is a subset of $GL(n, F)$, show that $[G, S]$ is connected.

2. If in Exercise 1 we have the additional conditions that G normalizes S and S is finite, show that $[G, S] = 1$ and so G centralizes S.

3. Show that a solvable connected subgroup of $GL(n, F)$ has solvable length ℓ where $2^{\ell-2} \leqslant n$.

4. Let G be the infinite dihedral group generated by

$$\begin{bmatrix} 0 & 1 \\ 1 & 0 \end{bmatrix} \quad \text{and} \quad \begin{bmatrix} a & 0 \\ 0 & a^{-1} \end{bmatrix}$$

in $GL(n, F)$ (where a is not a root of 1). Show that $\bigcap_{i=1}^{\infty} \gamma_i(G) = 1$; but conclude from Lemma 8.2B that for some integer $m > 0$, $G \cap \overline{\gamma_m(G)} = G \cap \overline{\gamma_{m+i}(G)} \neq 1$ for all $i \geqslant 0$.

5. Let G be a completely reducible solvable subgroup of $GL(n, F)$, and suppose that either char $F = 0$ or char $F \nmid |G\colon G_0|$ where G_0 is the connected component of 1. Show that each element in G is semisimple. [*Hint:* See Theorem 5.6.]

§8.7 We recall that a matrix $x \in GL(n, F)$ is *semisimple* if it is conjugate to a diagonal matrix in $GL(n, F)$ (F is algebraically closed); and x is *unipotent* if all its eigenvalues equal 1, so $(x - 1)^n = 0$.

Note. The unipotent matrices in $GL(n, F)$ constitute an algebraic set but the semisimple matrices do not.

Let $\eta \in F$, $\eta \neq 0$ and let

$$y = \begin{bmatrix} \eta & 0 & \cdots & 0 \\ 1 & \eta & \cdots & 0 \\ 0 & 1 & \ddots & \vdots \\ \vdots & \vdots & \ddots & \vdots \\ 0 & 0 & \cdots 1 & \eta \end{bmatrix}$$

be a matrix of the kind which appears as a block in the Jordan canonical form. Then $y_s = \operatorname{diag}(\eta, \eta, \ldots, \eta)$ and $y_u = yy_s^{-1}$ are semisimple and unipotent matrices, respectively, and $y_u y_s = y_s y_u$. Since each $x \in GL(n, F)$ is conjugate to its Jordan canonical form, we can always find x_s, $x_u \in GL(n, F)$ which are semisimple and unipotent, respectively, such that $x = x_u x_s = x_x x_u$. We now assert that x_u and x_s are uniquely determined by the conditions:

$$x = x_u x_s = x_s x_u \quad \text{and} \quad x_s \text{ is semisimple} \\ \text{and } x_u \text{ is unipotent.} \tag{8.7.1}$$

A direct proof of this is as follows. Let V be the underlying vector space of $GL(n, F)$, and let $\xi_1, \ldots, \xi_k$ be the different eigenvalues of x, and put

$$V_i = \{v \in V | v(x - \xi_i 1)^n = 0\} = \ker(x - \xi_i 1)^n.$$

Then it is easily shown that $V = \bigoplus_{i=1}^{k} V_i$ and that $V_i y \subseteq V_i$ for each i for each matrix y which commutes with x. Let x_s and x_u satisfy (8.7.1). Then $V_i x_s = V_i = V_i x_u$ for each i; since the restriction $x|V_i$ has the single eigenvalue ξ_i, we have $x_s|V_i = \xi_i 1$ and $x_u|V_i = \xi_i^{-1} x|V_i$. Thus the restrictions of x_s and x_u on each V_i are determined uniquely by x. Hence x_s and x_u are uniquely determined on V.

DEFINITION. For each $x \in GL(n, F)$, the matrices x_s and x_u given in (8.7.1) are called the *semisimple* and *unipotent* parts of x, respectively. We call $x = x_s u_u$ the *Jordan decomposition*.

Our interest in the Jordan decomposition is explained by the theorem to be proved in the next section.

Exercise. Let char $F = 0$. If $x \in GL(n, F)$ is unipotent, show that the cyclic group $\langle x \rangle$ is connected. What about the case where char $F = p > 0$? [*Hint:* Prove that $k \mapsto x^k$ is a continuous mapping (in the Zariski topologies) from the copy **Z** of the integers in F to

$GL(n, F)$. Then use Exercise 1 of §8.3 to show that **Z** is connected.]

§8.8 The object of this section is to prove the following theorem of E. R. Kolchin (1948).

THEOREM 8.8. *Let G be an algebraic group in* $GL(n, F)$. *If* $x \in G$ *has the Jordan decomposition* $x = x_s x_u$, *then* $x_s, x_u \in G$.

Clearly the result may be false if G is not algebraic, so it is not surprising that the proof of the theorem requires an analysis of sets of zeros of certain families of polynomials. The cases where char $F = 0$ *and where* char $F = p > 0$ *differ and must be dealt with separately. We shall give the main parts of the proof in a series of lemmas.*

LEMMA 8.8A. *Let F be of characteristic 0 (but not necessarily algebraically closed). Let* $f(X_0, X_1, \ldots, X_n) \neq 0$ *be a polynomial over F such that, for some* $x = (\xi_1, \ldots, \xi_n)$ *with each* ξ_i *a non-zero element of F, we have*

$$f(j, \xi_1^j, \ldots, \xi_n^j) = 0 \quad \text{for all sufficiently large integers } j \geq 0. \qquad (8.8.1)$$

Then $f(X_0, 1, \ldots, 1)$ *is identically zero.*

Proof. Write $f(X_0, \ldots, X_n) = \sum_{i=1}^{k} p_i(X_0) m_i(X)$ where the $m_i(X) = X_1^{i_1} \ldots X_n^{i_n}$ are distinct monomials and each $p_i(X_0) \neq 0$. We have to prove that $\sum_{i=1}^{k} p_i(X_0)$ is identically zero. By an induction argument we may suppose that if

$$g(X_0, \ldots, X_n) = \sum_{i=1}^{k'} p_i'(X_0) m_i'(X)$$

is a polynomial of the same form with either $k' < k$ or $k' = k$ and $\Sigma \deg p_i'(X_0) < \Sigma \deg p_i'(X_0)$ and if $g(X_0, \ldots, X_n)$ satisfies a condition like (8.8.1), then $g(X_0, 1, \ldots, 1)$ is identically zero. However, for any integer ℓ, $1 \leqslant \ell \leqslant k$, the polynomial

$$g(X_0, \ldots, X_n) = m_\ell(x)\, f(X_0, \ldots, X_n) - f(X_0+1, \xi_1 X_1, \ldots \xi_n X_n)$$
$$= \sum_{i=1}^{k} \{m_\ell(x) p_i(X_0) - m_i(x) p_i(X_0+1)\}\, m_i(X)$$

satisfies (8.8.1) in place of f. Therefore we conclude

$$g(X_0, 1, \ldots, 1) = \sum_{i=1}^{k} \{m_\ell(x)\, p_i(X_0) - m_i(x) p_i(X_0+1)\} = 0,$$

and so

$$m_\ell(x) \sum_{i=1}^{k} p_i(X_0) = \sum_{i=1}^{k} m_i(x) p_i(X_0 + 1). \tag{8.8.2}$$

This is true for all ℓ, $1 \leqslant \ell \leqslant k$. Since the right-hand side of (8.8.2) is independent of ℓ, we conclude $\sum_{i=1}^{k} p_i(X_0) = 0$, except in the case where all $m_\ell(x)$ are equal. However, in the latter case,

$$f(j, \xi_1^j, \ldots, \xi_n^j) = m_\ell(x)^j \sum_{i=0}^{k} p_i(j),$$

and so $\sum_{i=1}^{k} p_i(j) = 0$ for all sufficently large integers $j \geqslant 0$ by (8.8.1). Since char $F = 0$, this last condition implies $\sum_{i=1}^{k} p_i(X_0)$ has infinitely many zeros, and so $\sum_{i=1}^{k} p_i(X_0) = 0$ in this case too. ■

LEMMA 8.8B. *Let G be an algebraic group contained in* Diag(n, F). *Let I be the ideal in $F[X] = F[X_1, \ldots, X_n]$ consisting* *of all* $f(X)$ *such that* $f(\xi_1, \ldots, \xi_n) = 0$ *whenever* $\mathrm{diag}(\xi_1, \ldots, \xi_n) \in G$. *Then the ideal* I *has a basis which consists of polynomials of the form* $g(X) = X_1^{i_1} \ldots X_n^{i_n} - X_1^{i_1} \ldots X_n^{j_n}$; *in particular, the coefficients of* $g(X)$ *lie in the prime subfield of* F.

Proof. Let J be the ideal generated by all polynomials of the form of $g(X)$ in I; we must prove $I \subseteq J$. Let $f(X) \in I$ and write $f(X) = \sum_{i=1}^{k} a_i m_i(X)$ where the $a_i \in F$ and the $m_i(X)$ are monomials. We proceed by induction on k to show $f(X) \in J$. The result is true if $k = 0$, so suppose $k > 1$.

Because G is a group,

$$f(\xi_1 X_1, \ldots, \xi_n X_n) = \sum_{i=1}^{k} a_i m_i(\xi_1, \ldots, \xi_n) m_i(X) \in I$$

for each $\mathrm{diag}(\xi_1, \ldots, \xi_n) \in G$. Hence, for each ℓ, $1 \leqslant \ell \leqslant k$,

$$\begin{aligned} m_\ell(\xi_1, \ldots \xi_n) f(X) - f(\xi_1 X_1, \ldots, \xi_n X_n) \\ = \sum_{i \neq \ell} a_i \{m_\ell(\xi_1, \ldots, \xi_n) - m_i(\xi_1, \ldots, \xi_n)\} m_i(X) \end{aligned}$$

is in J by the induction hypothesis. Thus

$$\{m_\ell(\xi_1, \ldots, \xi_n) - m_s(\xi_1, \ldots, \xi_n\} f(X) \in J \quad \text{for all } \ell \text{ and } s.$$

This immediately implies $f(X) \in J$ unless we have $m_\ell(\xi_1, \ldots, \xi_n) = m_s(\xi_1, \ldots, \xi_n)$ for all ℓ and s. But in the latter case

$$\left(\sum_{i=1}^{k} a_i\right) m_1(\xi_1, \ldots, \xi_n) = f(\xi_1, \ldots, \xi_n) = 0$$

and so $\sum_{i=1}^{k} a_i = 0$. Thus induction again shows that

$$f(X) = \sum_{i=1}^{k} a_i m_i(X) - (\sum_{i=1}^{k} a_i) m_1(X)$$
$$= \sum_{i=2}^{k} (a_i - a_1)\, m_i(X) \in J.$$

Hence for all $f(X) \in I$ we have $f(X) \in J$, and the proof of the lemma is finished. ■

LEMMA 8.8C. *Let* $G \subseteq \mathrm{Diag}(n, F)$ *be an algebraic group. If* char $F = p > 0$, *then* $x = \mathrm{diag}(\xi_1, \ldots, \xi_n) \in G$ *whenever* $x^p \in G$.

Remark. This shows that for any semisimple $x \in GL(n, F)$, x is contained in the closure of $\langle x^p \rangle$ when char $F = p$.

Proof. It is enough to show that for each $f(X) = f(X_1, \ldots, X_n) \in F[X]$ such that $f(x^p) = 0$ we also have $f(x) = 0$. By Lemma 8.8B it is enough to do this in the case where $f(X)$ has its coefficients in the prime subfield of F. But for each λ in the prime subfield of F, $\lambda^p = \lambda$, and so $f(X_1, \ldots, X_n)^p = f(X_1^p, \ldots, X_n^p)$ (Compare with (5.5.1)). Therefore $f(x^p) = 0$ implies $f(x) = 0$ as required. ■

Proof of Theorem 8.8. We may suppose that G is the closure of the cyclic group $\langle x \rangle$ in $GL(n, F)$. Since $y \mapsto u^{-1}yu$ is a homeomorphism of $GL(n, F)$ we may also suppose that x is in its Jordan canonical form: $x = x_s x_u$ where $x_s = \mathrm{diag}(\xi_1, \ldots, \xi_n)$ is the semisimple part and $z = x_u - 1$ is nilpotent. Then, for each integer $j \geqslant 0$,

$$x^j = \mathrm{diag}(\xi_1^j, \ldots, \xi_n^j) \sum_{i=1}^{n} \binom{j}{i} z^i \tag{8.8.3}$$

where the binomial coefficients $\binom{j}{i}$ are polynomials in j (for fixed i) whenever char $F = 0$. We consider the cases char $F = 0$ and char $F = p > 0$ separately.

Case 1. (char $F = 0$). From (8.8.3) any polynomial in the entries of x^j is a polynomial in $j, \xi_1^j, \ldots, \xi_n^j$. But Lemma 8.8A shows that $f(j, \xi_1^j, \ldots, \xi_n^j) = 0$ for all $j \geqslant 0$ implies $f(1, 1, \ldots, 1) = 0$. However, when we substitute 1 for each of $j, \xi_1^j, \ldots, \xi_n^j$ in the right-hand side of (8.8.3) we get $1 + z = x_u$. Therefore for each polynomial $g(X)$ such that $g(x^j) = 0$ for all $j \geqslant 0$ we have $g(x_u) = 0$. This shows that $x_u \in G$, and hence also $x_s = xx_u^{-1} \in G$.

Case 2. (char $F = p > 0$). Since $\binom{p^m}{i} \equiv 0 \pmod{p}$ for $i = 1, \ldots, p^m - 1$, (8.8.3) shows that

$$x^{p^m} = \mathrm{diag}(\xi_1^{p^m}, \ldots, \xi_n^{p^m}) \in G \cap \mathrm{Diag}(n, F)$$

whenever $p^m > n$. Because $\mathrm{Diag}(n, F)$ is an algebraic group, so is $G \cap \mathrm{Diag}(n, F)$. Therefore $x_s = \mathrm{diag}(\xi_1, \ldots, \xi_n) \in G$ by Lemma 8.8C, and hence also $x_u = xx_s^{-1} \in G$. ∎

Exercises. 1. Show that the binomial coefficient $\binom{j}{i}$ (for fixed i) is not a polynomial function in j over a field of characteristic p where $p \leqslant i$.

2. If $G \subseteq GL(n, F)$ is an abelian algebraic group, show that $G = G_u \times G_s$ where G_u and G_s are the subsets of G consisting of all unipotent and all semisimple elements, respectively.

3. Use Lemma 8.8C to show that if G is a completely reducible subgroup of $GL(n, F)$, A a self-centralizing normal abelian subgroup of G, and char $F = p > 0$, then G/A contains no non-trivial p-element. (Compare with the proof of Theorem 6.6.)

§8.9 *Notes and References*

We begin with two general references to the material of this Chapter.

Borel, A. Groupes linéaires algebriques, *Ann. Math* (2) **64**, 20–82 (1956).

Kaplansky, I. *An Introduction to Differential Algebra.* Hermann, Paris (1957).

Most of the material is based on these two sources.

The earliest study of algebraic groups (mostly for characteristic 0) is found in

Chevalley, C. *Théorie des Groupes de Lie,* Tome II, Paris (1951).

Some earlier isolated results appear in two papers of Kolchin, namely the paper quoted in §2.11 and

Kolchin, E. R. On certain concepts in the theory of matrix groups, *Ann. Math.* (2) **49**, 774–789 (1948).

Our treatment of Theorem 8.8 is based in part on Kolchin's early proof.

By using deeper methods of algebraic geometry, extensive results have been obtained in the theory of algebraic groups. See

Borel, A. and Mostow, G. D. (eds.), Algebraic Groups and Discontinuous Subgroups, *Proc. Symposia in Pure Math.* Vol. 9, Amer. Math Soc., Rhode Island (1966).

On the other hand, the following paper is highly recommended as an introduction to applications of the methods of algebraic groups to the structure of linear groups.

Platonov, V. P. Linear and periodic groups, *Amer. math. Soc. Transl.* (2) **69**, 61–110 (1968).

For example, in this last paper Platonov proves that the Sylow p-groups of an algebraic group are always conjugate, and that an algebraic group in which each finite subgroup is solvable is solvable.

CHAPTER 9

Periodic Linear Groups

§9.1 One part of the theory of infinite groups deals with the ways in which various 'finiteness conditions' on groups are related. In this chapter we examine this question with respect to linear groups.

The classical theory of Schur in §9.2 shows that one of the weakest finiteness conditions, namely periodicity, is equivalent to local finiteness. Then in §9.3 we have a recent result of D. J. Winter which shows that a completely reducible periodic linear group is countable. Theorem 9.4 shows that a completely reducible linear p-group satisfies the minimal condition on subgroups. In §9.5 we extend results on finite groups to show that a non-modular periodic linear group is 'almost abelian'. Finally, in Theorem 9.6 we apply several of these results to prove that the Sylow p-groups of a periodic linear group are always conjugate.

§9.2 A locally finite group (one in which each finite subset generates a finite subgroup) is clearly periodic. In general the converse is false (see the Remark below), but the theorem of this section shows that the converse is true for linear groups. The proof of this result is essentially due to Schur (1911) with modifications by Kaplansky (1957) to include the case where the field has non-zero characteristic. We need the following general lemmas.

LEMMA 9.2A. *Let G be a finitely generated group. Then each subgroup H of finite index in G is finitely generated.*

Proof. Choose $x_1, \ldots, x_n \in G$ so that each element in G is equal to a product of x_i's; this can be done by taking a set of generators for G together with the inverses of these generators. Let $u_1 = 1, \ldots, u_h$ be a set of right coset representatives for H in G, and define $y_{ij} \in H$ by $u_i x_j = y_{ij} u_k$ for some k ($i = 1, \ldots, h$; $j = 1, \ldots, n$). We shall show that the y_{ij} generate H. Indeed, each $y \in H$ has the form $y = x_{j_1} \ldots x_{j_m}$ so

$$y = u_1 y = y_{1j_1} u_{k_2} (x_{j_2} \ldots x_{j_m})$$

$$= \ldots = y_{k_1 j_1} \ldots y_{k_m j_m} u_{k_{m+1}}$$

where $k_1 = 1, k_2, \ldots, k_{m+1}$ are defined successively by $u_{k_s} x_{j_s} = y_{k_s j_s} u_{k_{s+1}}$ ($s = 1, \ldots, m$). Since $y \in H$, $u_{k_{m+1}} = 1$, and so $y = y_{k_1 j_1} \ldots y_{k_m j_m}$ as required. ∎

LEMMA 9.2B. *Let N be a normal subgroup of G. If N and G/N are both locally finite, then G is locally finite.*

Proof. Let H be a finitely generated subgroup of G. Since G/N is locally finite, the finitely generated group HN/N is finite. Hence $H/H \cap N$ is finite and so $H \cap N$ is finitely generated by Lemma 9.2A. Since N is locally finite, $H \cap N$ is finite, and so H is finite.

Note. 1. Clearly a periodic abelian group is locally finite. Therefore, by induction on solvable length, we can deduce from Lemma 9.2B that each solvable periodic group is locally finite.

2. In particular, if F is a field of characteristic $p > 0$, then $STL(n, F)$ is locally finite (see Lemma 1.3).

THEOREM 9.2. *A periodic linear group is locally finite.*

Proof. We have to show that if G is a finitely generated periodic subgroup of $GL(n, F)$, then G is finite. Clearly we may suppose F to be algebraically closed.

Let $x_1, \ldots, x_h$ generate G and let F_1 be the subfield of F generated by the entries of these matrices; clearly $G \subseteq GL(n, F_1)$. Since F_1 is finitely generated, there is a finite *transcendence basis* $\xi_1, \ldots, \xi_k$ of F_1 over the prime subfield F_0; that is, a maximal set of elements algebraically independent over F_0. Put $E = F_0(\xi_1, \ldots, \xi_k)$. Since F_1 is finitely generated by algebraic elements over E, $[F_1 : E] = d$, say, is finite. Now Example 3 of §3.2 shows that G is isomorphic to a subgroup H of $GL(nd, E)$. It remains to show that the finitely generated periodic group H is finite. We have two cases.

Case 1 (char $F = 0$). In this case F_0 is isomorphic to $\mathbf{Q}$. For each $x \in H$, let $f_x(X) \in E[X]$ denote the (monic) minimal polynomial for x; x has finite order so the zeros of $f_x(X)$ are all roots of 1. Thus the coefficients of $f_x(X)$ are sums of roots of 1 and hence algebraic integers. But, by the definition of E, the only elements of E algebraic over F_0 are in F_0. Thus the coefficients of $f_x(X)$ are rational integers. Since the coefficients of $(X+1)^{nd}$ dominate the coefficients of $f_x(X)$, we conclude that there are only a finite number of different $f_x(X)$ as x runs through H. Since x and y are of the same order when $f_x(X) = f_y(X)$, we conclude that H has a finite exponent. Therefore H is finite by Theorem 2.9.

Case 2 (char $F = p > 0$). In this case F_0 contains p elements. As in Case 1 we can show that $f_x(X) \in F_0[X]$ for all $x \in H$. Thus, once again there are only finitely many different $f_x(X)$ as x runs through H, and so H has finite exponent. Finally, Theorem 2.9 shows that H has a subgroup N of finite index such that N is conjugate to a subgroup $STL(nd, E)$. Since N is finitely generated by Lemma 9.2A, and $STL(nd, E)$ is locally finite (see the Note above), we conclude that N is finite; hence H is finite. ■

Remark. There was a long-outstanding question as to whether there is a p-group which is not locally finite. This was answered in the affirmative by Golod and Shaferevich in 1964. For example, see I. N. Herstein, Noncommutative Rings, *Carus Monograph,* Vol. 15. Wiley, New York (1968).

§9.3 The following result is due to David J. Winter (1968).

THEOREM 9.3. *Let F be an algebraically closed field and let F_1 be the algebraic closure of the prime subfield in F. If G is a completely reducible periodic subgroup of $GL(n, F)$ then G is conjugate to a subgroup of $GL(n, F_1)$.*

Remark. G is locally finite by Theorem 9.2.

Proof. Since G is assumed to be completely reducible it is enough to consider the case where G is irreducible. Choose $x_1, \ldots, x_s$ in G as a basis for FG over F, and put $H = \langle x_1, \ldots, x_s \rangle$. Since G is locally finite, H is finite; and since $FG = FH$ and G is irreducible, H is irreducible over F. Because F_1 contains each finite extension of its prime subfield, Theorem 3.4B shows that $a^{-1}Ha \subseteq GL(n, F_1)$ for some $a \in GL(n, F)$. We shall now prove that $a^{-1}ya \in GL(n, F_1)$ for each $y \in G$, and hence $a^{-1}Ga \subseteq GL(n, F_1)$.

For each $y \in G$, the subgroup $H_1 = \langle H, y \rangle$ is finite because G is locally finite. Therefore, as before, $b^{-1}H_1b \subseteq GL(n, F_1)$ for some $b \in GL(n, F)$. Thus we have two irreducible representations of H over F_1, namely $x \leftrightarrow a^{-1}xa$ and $x \leftrightarrow b^{-1}xb$, with the same character. These representations are equivalent (over F_1) by Theorem 2.7, so there exists $c \in GL(n, F_1)$ such that $c^{-1}(a^{-1}xa)c = b^{-1}xb$ for all $x \in H$. Now $b^{-1}yb \in GL(n, F_1)$ by the choice of b, and so

$a^{-1}ya \in GL(n, F_1)$ because $c \in GL(n, F_1)$. This is true for all $y \in G$; hence $a^{-1}Ga \subseteq GL(n, F_1)$. ■

COROLLARY 9.3. *Let G be a completely reducible periodic linear group. Then G is countable, and there is an increasing sequence of finite subgroups,* $1 = G_0 \subseteq G_1 \subseteq \ldots$, *such that* $G = \bigcup_{n=0}^{\infty} G_i$.

Proof. From Corollary 2.8C it is enough to consider the case where the underlying field F is algebraically closed. Using the notation of the Theorem we note that the algebraic closure F_1 of the prime subfield of F is always countable, and hence $GL(n, F_1)$ is countable; therefore it follows from the Theorem that G is countable. If we enumerate the elements of G in some order $1 = x_0, x_1, \ldots$, and define $G_i = \langle x_0, x_1, \ldots, x_i \rangle$ $(i = 0, 1, \ldots)$, then the G_i are finite subgroups of G because G is locally finite, and $G = \bigcup_{i=0}^{\infty} G_i$ as required. ■

Exercise. Show that both these results may fail to be true if we drop the condition that G is completely reducible. [*Hint:* take $G = STL(n, F)$.]

§9.4 This section examines the structure of linear p-groups in the case where the underlying field has characteristic $\neq p$.

DEFINITION. A group G satisfies the *minimal condition* (on subgroups) if each non-empty set of subgroups of G possesses an element which does not contain any other element of the set. Equivalently, every infinite sequence of subgroups of the form $H_1 \supseteq H_2 \supseteq \ldots$ has the property that for some integer m, $H_m = H_{m+i}$ for all $i \geqslant 0$.

Note. 1. If G satisfies the minimal condition then so do all subgroups and all quotient groups of G.

2. If G satisfies the minimal condition then G is periodic; indeed consider the sequence of subgroups $\langle x \rangle \supset \langle x^2 \rangle \supset \langle x^4 \rangle \supset \ldots$ of any infinite cyclic group $\langle x \rangle$. On the other hand, an infinite abelian group of finite exponent is periodic but does not satisfy the minimal condition.

We shall need the following general lemmas.

LEMMA 9.4A. *Let N be a normal subgroup of a group G. If N and G/N both satisfy the minimal condition, then so does G.*

Proof. Let Σ be a non-empty set of subgroups of G. Then the set $\Sigma_1 = \{SN/N \mid S \in \Sigma\}$ of subgroups of G/N has a minimal element, say TN/N. Moreover the set $\Sigma_2 = \{S \cap N \mid S \in \Sigma \text{ and } SN = TN\}$ of subgroups of N has a minimal element, say $M \cap N$, with $M \in \Sigma$. We claim M is a minimal element of Σ. For suppose $S \in \Sigma$ and $S \subseteq M$. Then $SN/N = TN/N = MN/N$ by the minimality of TN/N in Σ_1; hence $SN = MN$. Similarly $S \cap N = M \cap N$ by the choice of M. Therefore $S = S(M \cap N) = M \cap SN = M \cap MN = M$. Hence M is minimal in Σ. ∎

LEMMA 9.4B. *Let p be a prime, and let F be a field with* char $F \neq p$. *Then*

$$A_0 = \{\xi \in F \mid \xi^{p^m} = 1 \text{ for some integer } m \geqslant 0\}$$

is an (abelian) group satisfying the minimal condition; indeed every proper subgroup of A_0 is finite.

Remark. If F is algebraically closed, then A_0 is a *quasi-cyclic* (or Prüfer) p-group.

Proof. Each of the sets $A_n = \{\xi \in F \mid \xi^{p^n} = 1\}$ is a finite subgroup of the multiplicative group of F, and so is cyclic (see §0.

We note that $A_0 = \bigcup_{n=1}^{\infty} A_n$. Suppose that B is an infinite subgroup of A_0. Then B must contain elements of arbitrarily large order and so $B \supseteq A_n$ for infinitely many n. Since $A_1 \subsetneq A_2 \subsetneq \ldots$, $B \supseteq \bigcup_{n=1}^{\infty} A_n = A$. Thus we conclude that each proper subgroup of A is finite, and trivially this implies the minimal condition is satisfied. ■

Note. It follows from Corollary 2.5 and Theorem 2.8B that a linear p-group over a field F is completely reducible $\Longleftrightarrow$ char $F \neq p$.

THEOREM 9.4. *Let G be a linear p-group over a field F with char $F \neq p$. Then G satisfies the minimal condition.*

Proof. Clearly we can suppose F is algebraically closed. By Theorem 9.2 each finitely generated subgroup of G is a finite p-group, and so G is locally nilpotent. Hence by Corollary 4.6B G is a monomial group. This means in particular that G has a normal subgroup A of finite index in G which corresponds to a subgroup of Diag(n, F) over a suitable choice of the basis for the space on which G acts. Now Diag(n, F) is isomorphic to the direct product of n copies of the multiplicative group of F, and so A is isomorphic to a subgroup of $A_0 \times \ldots \times A_0$ (n times) where A_0 is defined in Lemma 9.4B. Thus we conclude from Lemmas 9.4B and 9.4A that A satisfies the minimal condition. Since G/A is finite, Lemma 9.4A then shows that G satisfies the minimal condition. ■

COROLLARY 9.4. *Let G be a linear p-group over a field F with char $F \neq p$, and let B be the intersection of all subgroups of finite index in G. Then:*

(1) *B is a normal abelian subgroup of finite index in G.*

(2) *B is a* divisible *abelian group; that is, for each $b \in B$ and each integer $h > 0$ there exists $a \in B$ such that $a^h = b$.*

(3) *B is the only divisible abelian subgroup of finite index in G.*

Remark. Clearly a divisible abelian group is finite only when it is 1. Hence G is finite $\Longleftrightarrow B = 1$.

Proof. Let B_1 be a minimal element in the set Σ of all subgroups of finite index in G. Since the intersection of any two subgroups of finite index in G is again of finite index in G, B_1 is contained in all the elements of Σ; therefore $B_1 = B$. Since G has an abelian subgroup of finite index (see the proof of the Theorem), B is abelian. Clearly B is normal in G, so (1) is proved. Now for each prime q define the subgroup $B_q = \{a^q \mid a \in B\}$. Then B/B_q is an elementary abelian q-group, so if $B \neq B_q$ there would be a subgroup of index q in B contrary to the definition of B. This shows $B = B_q$ for all primes q, and this implies (2). Finally, if D is a divisible abelian subgroup of finite index in G, then $D \supseteq B$ by the definition of B. If $D \neq B$, then p divides $|D/B|$ and so $D_p = \{a^p \mid a \in D\}$ is a proper subgroup of D; this contradicts the divisibility of D. Hence (3) is proved. ■

Exercises. 1. Find necessary and sufficient condition in order that a periodic abelian group A should have a faithful representation of degree n over a field F. [*Hint:* Consider the p-subgroups of A and separate out the cases char $F = p$ and char $F \neq p$.]

2. Let F be an infinite field with char $F = p > 0$. Show that the only linear p-groups over F which satisfy the minimal condition are finite. [*Hint:* Use Theorem 2.8B.]

§9.5 The present section extends the result of Theorem 5.7 from finite groups to periodic groups.

We shall extend the definition of non-modular from finite groups in the obvious way: a periodic linear group over a field F is *non-modular* if either char $F = 0$ or char $F = p > 0$ but no element of the group has order divisible by p.

Note 1. It follows from Theorem 9.2 and Corollary 2.5 that each non-modular periodic linear group is completely reducible. As we have seen, the converse is false, even for finite groups.

2. Each element x in a non-modular periodic linear group over an algebraically closed field is semisimple, because $\langle x \rangle$ is completely reducible.

THEOREM 9.5. *Let G be a non-modular periodic linear group of degree n over a field F. Then G has a normal abelian subgroup of index at most $\beta(n)$ where $\beta(n)$ is a bound depending only on n.*

Remark. Theorem 5.7 shows that $\beta(n) \leqslant (49n)^{n^2}$.

Proof. We may suppose F to be algebraically closed. By Note 1 above, G is completely reducible and so by Corollary 9.3 we have $G = \bigcup_{i=1}^{\infty} G_i$ where $G_1 \subsetneq G_2 \subsetneq \ldots$ is a sequence of finite subgroups of G.

Now let $\beta(n)$ be the bound such that every finite non-modular linear group has a normal abelian subgroup of index $\leqslant \beta(n)$ (see Theorem 5.7). Then the set Γ_i of all normal abelian subgroups of index $\leqslant \beta(n)$ in G_i is not empty ($i = 1, 2, \ldots$). We note that if $B \in \Gamma_i$, and $1 \leqslant j \leqslant i$, then $|G_i : B| \geqslant |G_jB : B| = |G_j : B \cap G_j|$ and so $B \cap G_j \in \Gamma_j$. In particular, each subgroup in $\Gamma = \bigcup_{i=1}^{\infty} \Gamma_i$ contains at least one subgroup in Γ_1. Since Γ_1 is finite we can choose $A_1 \in \Gamma_1$ so that A_1 is contained in infinitely many subgroups in Γ. In general, if we have chosen $A_j \in \Gamma_j$ for $j = 1, \ldots, i-1$ such that $A_1 \subsetneq A_2 \ldots \subsetneq A_{i-1}$ and A_{i-1} is contained in infinitely many subgroups in Γ, then A_{i-1} is contained in certain of the subgroups in Γ_i. Since Γ_i is finite we can choose $A_i \in \Gamma_i$ so that $A_{i-1} \subsetneq A_i$

and A_i is contained in infinitely many subgroups in Γ. In this way we can choose an infinite sequence $A_1 \subsetneq A_2 \subsetneq \ldots$ with $A_i \in \Gamma_i$, and we define $A = \bigcup_{i=1}^{\infty} A_i$. It is readily seen that A is a normal abelian subgroup of G. Moreover, if $s = \beta(n) + 1$, then for any s elements $x_1, \ldots, x_s \in G$ we have $x_1, \ldots, x_s \in G_m$ for sufficiently large m, and then $x_i x_j^{-1} \in A_m \subseteq A$ for some $i \neq j$ because $|G_m : A_m| < s$. This shows that A has at most $\beta(n)$ cosets, and so A is the required subgroup. ■

Exercises. 1. Let G be a non-modular periodic linear group. Show that if G is connected (in the Zariski topology), then G is abelian. [*Hint:* Use Theorem 8.5 and Lemma 8.6B.]

2. Let G be a non-modular periodic linear group of degree n. If p is a prime with $p > (2n+1)(n-1)$, show that the set of all p-elements of G form a normal subgroup of G. [*Hint:* use Theorem 5.5.]

3. Generalize Theorem 5.4 to periodic linear groups.

4. Let G be an algebraic group over an algebraically closed field of characteristic 0. If G is periodic, prove that G is finite.

5. Let F be the algebraic closure of a finite field. Show that $GL(n, F)$ is a periodic algebraic group.

§9.6 In any group G we can define a *Sylow p-group* to be any subgroup of G which is maximal in the set of p-subgroups of G. An easy application of Zorn's lemma shows that each p-subgroup of G is contained in a Sylow p-group. However, the well-known Sylow theorems may no longer hold when G is an infinite group, even if G is a linear group (see Exercise below). However, there are some classes of linear groups in which we can prove the conjugacy theorems for Sylow p-groups, and we shall give such a theorem for periodic linear groups. These results are due to V. P. Platonov (1966) and were proved independently by B. Wehrfritz.

We shall need the following general lemma.

LEMMA 9.6. *Let G be a locally finite group, and suppose P_1 and P_2 are Sylow p-groups of G and P_1 is finite. Then P_1 is conjugate to P_2 in G.*

Proof. For each finitely generated subgroup S of P_2, $H = \langle S, P_1 \rangle$ is a finite group, and so S is conjugate to a subgroup of P_1 by the Sylow theorems for finite groups. Hence we conclude $|P_2| \leqslant |P_1|$. Therefore P_1 and P_2 are Sylow p-groups of the finite group $\langle P_1, P_2 \rangle$ and hence are conjugate. ■

THEOREM 9.6. *Let G be a periodic subgroup of $GL(n, F)$ over any field F. If P_1 and P_2 are Sylow p-groups of G, then P_1 is conjugate to P_2.*

Proof. Clearly we may suppose that F is algebraically closed. We shall frequently use the fact that G is locally finite (Theorem 9.2). We consider two cases.

Case 1 (char $F = p$). Choose a finite subset S_i of P_i so that $FS_i = FP_i$ $(i = 1, 2)$. Then $H = \langle S_1, S_2 \rangle$ is a finite group, and we can choose a Sylow p-group Q_i of H such that $S_i \subseteq Q_i$ for $i = 1, 2$. Since each p-element of $GL(n, F)$ is unipotent (because char $F = p$), Theorem 2.8B shows that $z_1^{-1}Q_1z_1 \subseteq STL(n, F)$ for some $z_1 \in GL(n, F)$. Since Q_1 and Q_2 are conjugate in H, there exists $z_2 \in GL(n, F)$ such that $z_2^{-1}Q_2z_2 \subset STL(n, F)$ and $z_1z_2^{-1} \in H$. But $FP_i = FS_i$, and so $z_i^{-1}P_iz_i \subseteq TL(n, F)$ for $i = 1, 2$. Since $STL(n, F)$

is the unique Sylow p-group of $TL(n, F)$, $z_i^{-1}Pz_i \subseteq STL(n, F)$ for each i. Thus $<z_1^{-1}P_1z_1, z_2^{-1}P_2z_2>$ is a p-subgroup of $z_1^{-1}Gz_1$ because $z_1z_2^{-1} \in H \subseteq G$. By the maximality of the Sylow p-groups we conclude that $z_1^{-1}P_1z_1 = z_2^{-1}P_2z_2$. Hence $P_2 = (z_1z_2^{-1})^{-1}P_1z_1z_2^{-1}$ with $z_1z_2^{-1} \in G$ as required.

Case 2 (char $F \neq p$). Let B_i be the normal divisible abelian subgroup of P_i defined in Corollary 9.4. Because B_i is divisible we can choose a finite subset $S_i \subset B_i$ such that the $(n!)$-powers of the elements in S_i form a set T_i with $FT_i = FB_i$ for $i = 1, 2$. Then $H = <S_1, S_2>$ is a finite group, and so by the Sylow theorems there exists $x \in H$ such that $K = <x^{-1}S_1x, S_2>$ is a (finite) p-group. Since K is conjugate in $GL(n, F)$ to a group of monomial matrices by Theorem 4.6B, K has a normal abelian subgroup A, say, whose index $|K : A|$ divides $n!$. By the choice of S_1 and S_2, both $x^{-1}T_1x$ and T_2 are therefore in A; hence $[x^{-1}T_1x, T_2] = 1$. But $FT_i = FB_i$ $(i = 1, 2)$, and so $[x^{-1}B_1x, B_2] = 1$. Now $N_G(B_2)/B_2$ is a locally finite group because G is, and it has a finite Sylow p-group, namely P_2/B_2. Since $B_2x^{-1}B_1x/B_2$ is a p-group it is contained in a Sylow p-group of $N_G(B_2)/B_2$, and so by Lemma 9.6 some conjugate $y^{-1}B_1y$ of $x^{-1}B_1x$ in $N_G(B_2)$ is contained in P_2. Again, the locally finite group $N_G(y^{-1}B_1y)/y^{-1}B_1y$ has a finite Sylow p-group $y^{-1}P_1y/y^{-1}B_1y$ and contains the p-subgroup $y^{-1}B_1yB_2/y^{-1}B_1y$. Therefore $|y^{-1}B_1yB_2 : y^{-1}B_1y|$ is finite by Lemma 9.6. Thus $y^{-1}B_1y \cap B_2$ is a subgroup of finite index in B_2 and hence equals B_2 by Corollary 9.4; therefore $y^{-1}B_1y \supseteq B_2$. Since $y^{-1}B_1y$ is a divisible abelian group, another application of Corollary 9.4 shows that $y^{-1}B_1y = B_2$. Finally, since $y^{-1}P_1y/B_2$ and P_2/B_2 are finite Sylow p-groups of $N_G(B_2)/B_2$, Lemma 9.6 shows that some conjugate of P_1 is equal to P_2. ■

Exercise. Let G be the subgroup of $GL(n, \mathbf{Q})$ generated by

$$\begin{bmatrix} 0 & 1 \\ 1 & 0 \end{bmatrix} \quad \text{and} \quad \begin{bmatrix} \alpha & 0 \\ 0 & \alpha^{-1} \end{bmatrix}$$

for all $\alpha \in \mathbf{Q}$, $\alpha \neq 0$. Show that G has infinitely many pairwise non-conjugate Sylow 2-groups (all of order 4).

§9.7 *Notes and References*

Our proof of Theorem 9.2 follows that of Kaplansky (see the reference given in §2.11).

Theorem 9.3 appears in

Winter, D. J. Representations of locally finite groups, *Bull. Am. math. Soc.* **74**, 145–148 (1968).

The results of §9.4 are due to Mal'cev; I first learned them from K. Hirsch.

Theorem 9.5 is classical in the case of the complex field and is due to Schur (1911). The proof given here is a modification of one due to G. Glauberman.

Theorem 9.6 is one of a series of generalizations of theorems true for finite groups to periodic linear groups. This work has been done independently by Platonov and Wehrfritz, and we shall give some more of Platonov's work in the next chapter. For Platonov's work see the reference given in §8.9 and the bibliography he gives there. The papers of Wehrfritz include

Wehrfritz, B. A. F. Sylow theorems for periodic linear groups, *Proc. Lond. math. Soc.* (3) **18**, 125–140 (1968).

——, Conjugacy theorems in locally finite groups II, *Arch. Math.* **18**, 470–473, (1967).

CHAPTER 10

The Method of Finite Approximation

§10.1 One way of looking at the fact that a group is a linear group is to view it as a finiteness condition on the group. In this final chapter we examine the condition of linearity from this viewpoint, and show how far we can extend properties true for finite groups. The basic method is that of 'finite approximation' which was first used by Mal'cev in 1940 and recently sharpened by Platonov.

In §10.2 we prove that a finitely generated linear group is residually finite in a strong sense. Theorem 10.3 and 10.4 show how these results may be used to give criteria for solvability and nilpotence, respectively, of linear groups. §10.5 gives a number of applications.

§10.2 The theorem of this section is a theorem of Platonov (1966) based on an earlier result of Mal'cev (1940).

LEMMA 10.2. *Suppose that for each prime q we choose an infinite field $E^{(q)}$ of characteristic q. Let $F = F_0(X_1, \ldots, X_k) = F_0(X)$ be the field of rational functions in k indeterminates over a prime subfield F_0 of arbitrary characteristic. Let G be a finitely generated subgroup of $GL(m, F)$. Then:*

(1) *If* char $F = q > 0$, *then for each $z \neq 1$ in G there is a homomorphism $\psi_1 : G \to GL(m, E^{(q)})$ for which $z^{\psi_1} \neq 1$.*

(2) *If* char $F = 0$, *then for each $z \neq 1$ in G there is a bound q_0*

(depending on z) such that for each prime $q \geqslant q_0$ there is a homomorphism $\psi_2 : G \to GL(m, E^{(q)})$ for which $z^{\psi_2} \neq 1$.

Proof. By taking a set of generators of G together with their inverses we can find $x_1, \ldots, x_h \in G$ such that each element in G is a product of x_i's.

Case 1 (char $F = q > 0$). Since $E^{(q)}$ is infinite we can find $\alpha_1, \ldots, \alpha_k \in E^{(q)}$ such that $a = (\alpha_1, \ldots, \alpha_k)$ is not a zero of any of the (finite set of) polynomials in $F_0[X]$ which occure as numerators or denominators in the non-zero entries of the matrices $x_1, \ldots, x_h, z - 1$. Then $f(X) \mapsto f(a)$ is a homomorphism of the ring $F_0[X]$ into $E^{(q)}$, and this homomorphism induces a homomorphism $\psi_1 : G \to GL(m, E^{(q)})$ (see §3.2) and $z^{\psi_1} \neq 1$ by the choice of $(\alpha_1, \ldots, \alpha_k)$.

Case 2 (char $F = 0$). In this case two steps are involved. We can take $F_0 = \mathbf{Q}$ and observe that each element in F is a quotient of elements in $\mathbf{Z}[X]$. We choose q_0 so that none of the polynomials in $\mathbf{Z}[X]$ which occur as numerators or denominators in the non-zero entries of the matrices $x_1, \ldots, x_h, z-1$ is divisible by a prime $q \geqslant q_0$. Then for each prime $q \geqslant q_0$, the mapping (mod q) induces a homomorphism $\psi_0 : G \to GL(m, \mathbf{Z}_q(X))$ where $\mathbf{Z}_q$ is the field of integers (mod q); note that $z^{\psi_0} \neq 1$ by the choice of q_0. But, considering G^{ψ_0} and z^{ψ_0}, we are now in Case 1. Therefore the composite of ψ_0 with the mapping ψ_1 as defined in Case 1 gives a homomorphism $\psi_2 = \psi_0 \circ \psi_1 : G \to GL(m, E^{(q)})$ with $z^{\psi_2} \neq 1$ as required. ■

We now prove the rather technical 'finite approximation' theorem. It is from this theorem that all our other results will follow.

THEOREM 10.2. *Suppose that for each prime q we choose an infinite field $E^{(q)}$ of characteristic q. If G is a finitely generated* *subgroup of* $GL(n, F)$ *where* F *is an arbitrary field, then there is an integer* $m > 0$ *(depending on* n *and* F *but independent of the choice of* $E^{(q)}$*) such that:*

(1) *If* char $F = q > 0$, *then for each* $z \neq 1$ *in* G *there is a homomorphism* $\psi : G \to GL(m, E^{(q)})$ *for which* $z^{\psi} \neq 1$.

(2) *If* char $F = 0$, *then for each* $z \neq 1$ *in* G *there is a bound* q_0 *such that for each prime* $q \geqslant q_0$ *there is a homomorphism* $\psi : G \to GL(m, E^{(q)})$ *for which* $z^{\psi} \neq 1$.

Proof. Let $y_1, \ldots, y_s$ be a set of generators for G and let F_1 be the subfield of F generated by the (finite) set of entries of the y_i. Then F_1 is finitely generated over the prime subfield F_0 and there is a transendence basis $\xi_1, \ldots, \xi_k$ for F_1 over F_0. If we put $F_2 = F_0(\xi_1, \ldots, \xi_k)$ then F_1 is infinitely generated and algebraic over F_2 and so $[F_1 : F_2] = d$, say, is finite. (Compare with the proof of Theorem 9.2.) Now $G \subseteq GL(n, F_1)$ and so Example 3 of §3.2 shows that there is an isomorphism θ of G onto a subgroup of $GL(n, F_2)$ where $m = nd$. But $F_2 = F_0(\xi_1, \ldots, \xi_k) \simeq F_0(X_1, \ldots, X_k)$, and so the theorem follows when we apply Lemma 10.2 to the group G^{θ}. ■

A group G is called *residually finite* if for each $z \neq 1$ in G there is a homomorphism of G onto a finite group for which the image of z is not 1; equivalently, G has a normal subgroup N_z of finite index such that $z \notin N_z$. If G is residually finite, then there is an isomorphism of G onto a subgroup of the unrestricted direct product $\bigotimes_z G/N_z$ of finite groups G/N_z $(z \in G, z \neq 1)$ given by $x \mapsto (N_z x)_z$.

Theorem 10.2 shows that a finitely generated linear group is residually finite in the following rather strong sense.

COROLLARY 10.2. *Let* F *be an arbitrary field, and let* G *be a finitely generated subgroup of* $GL(n, F)$. *Then there is an integer*

$m > 0$ such that for each $z \neq 1$ in G there is a finite field F_z and a homomorphism $\psi_z : G \to GL(m, F_z)$ for which $z^{\psi_z} \neq 1$. Moreover, if char $F = q > 0$, *then we may take* char $F_z = q$ *for each* z.

In particular, no finitely generated infinite linear group is simple.

Proof. Choose each field $E^{(q)}$ in the Theorem as the algebraic closure of its prime subfield. Now define $\psi_z = \psi$ according to the Theorem. Since G^{ψ} is finitely generated, the subfield F_z of $E^{(q)}$ generated by the entries of the matrices in G^{ψ} is a finitely generated algebraic extension of the prime subfield. Hence F_z is a finite field and $\psi : G \to GL(m, F_z)$ as required. ■

Exercise. Show that we may take $m = n$ in Corollary 10.2. [*Hint:* Give a new proof using Theorem 3.3B. Note that $p(\xi) \neq 0 \iff p(\xi)Y - 1$ has a root.]

§10.3 In this section we use Corollary 10.2 to obtain a solvability criterion for certain classes of linear groups.

THEOREM 10.3. *Let Γ be a class of groups such that*

(1) *if G is in Γ, then each finitely generated subgroup and each finite homomorphic image of G is in Γ;*

(2) *if G is in Γ and G is finite, then G is solvable.*

Then every linear group in Γ is solvable.

Proof. Let G be a linear group of degree n and suppose G is in Γ. Let H be a finitely generated subgroup of G; then H is in Γ by (1). Moreover, by Corollary 10.2 and note preceding it, H is isomorphic to a subgroup of an unrestricted direct product $H^* = \bigotimes_{\lambda} H_\lambda$ where each H_λ is a linear group of degree m over a finite field (so H_λ is finite), and H_λ is a homomorphic image of H. Then each H_λ is in Γ by (1) and each H_λ is solvable by (2). Therefore, by Theorem 6.2A, $H_\lambda^{(2m)} = 1$ for all λ. This implies that $(H^*)^{(2m)} = 1$

and so $H^{(2m)} = 1$. Hence each finitely generated subgroup of G is solvable, so G is locally solvable. Theorem 6.2B now shows that G is solvable. ■

Note. If condition (1) is weakened to:

'(1)′ if G is in Γ, then each finite homomorphic image of G is in Γ;'

then we can conclude that each finitely generated linear group in Γ is solvable (take $H = G$ in the proof above).

Exercises. 1. It is known that a finite group in which each proper subgroup is nilpotent is solvable (see Huppert's book p. 280). Show that a linear group in which each proper subgroup is locally nilpotent is solvable.

2. It is known that a finite group in which each Sylow subgroup is cyclic is solvable of length $\leqslant 2$ (see Huppert's book p. 420). Show that a periodic linear group with each Sylow subgroup cyclic is solvable of length $\leqslant 2$.

3. It is known that any finite group which is a product of two nilpotent subgroups is solvable (see Huppert's book p.674). Show that a finitely generated linear group which is a product of two locally nilpotent groups is solvable.

§10.4 The theorem of this section shows how we may apply Theorem 10.2 to extend nilpotency criteria in finite groups to analogous criteria in linear groups. In this case we have to choose the fields $E^{(q)}$ mentioned in Theorem 10.2 rather carefully.

LEMMA 10.4A. *Let q be a prime, and let m be an integer > 0. Let $r > m$ be a prime. Suppose E_0 is a field of q elements, and define E_k ($k = 1, 2, \ldots$) successively so that E_k is an extension* *field of degree r over* E_{k-1}. *Put* $E^{(q)} = \bigcup_{k=0}^{\infty} E_k$.

(1) $E^{(q)}$ *is an infinite algebraic extension of its prime subfield* E_0.

(2) *If* $p \leqslant m$ *is a prime different from q, then each finite p-subgroup of* $GL(m, E^{(q)})$ *has order dividing* $\prod_{i=1}^{m} (q^i - 1)$.

Proof. The proof of (1) is immediate, so consider (2). Let P be a finite p-subgroup of $GL(m, E^{(q)})$. Since P is finite, the entries in the matrices in P all lie in E_k for some $k \geqslant 1$. Since E_k has q^{r^k} elements, Theorem 1.2 shows that

$$|P| \text{ divides } \prod_{i=1}^{m} (q^{ir^k} - 1). \tag{10.4.1}$$

Now for any integer $j \geqslant 1$ the greatest common divisor of the Euler function $\varphi(p^j)$ and r^k is 1 because $p < r$; therefore $a\varphi(p^j) + br^k = 1$ for some integers a, b. Moreover $q^{\varphi(p^j)} \equiv 1 \pmod{p^j}$ by Fermat's theorem. Thus, if $q^{ir^k} \equiv 1 \pmod{p^j}$, then $q^i = q^{ia\varphi(p^j)} q^{ibr^k} \equiv 1 \pmod{p^j}$. Hence we conclude from (10.4.1) that $|P|$ divides $\prod_{i=1}^{m} (q^i - 1)$ as required. ■

LEMMA 10.4B. *Let p and q be distinct primes and suppose* $m \geqslant 1$ *is an integer such that* $p \leqslant m$ *and* $q = 1 + \ell p^m$ *with* $p \nmid \ell$. *Then* p^{2m^2} *does not divide* $\prod_{i=1}^{m} (q^i - 1)$.

Proof. It is enough to show that for each i, $1 \leqslant i \leqslant m$, $p^{2m} \nmid q^i - 1$. Indeed, by the binomial theorem, $q^i - 1 = (1+\ell p^m)^i - 1 \equiv i\ell p^m \pmod{p^{2m}}$. Since $i \leqslant m < p^m$, $p^m \nmid i\ell$, and so $p^{2m} \nmid q^i - 1$. ■

THEOREM 10.4. *Let G be a finitely generated linear group of degree n over a field F, and let H be a subgroup of G. Suppose that for each normal subgroup N of finite index in G, the group HN/N is nilpotent. Then H is nilpotent.*

Proof. We have two cases.

Case 1 (char $F = q > 0$). By Theorem 10.2, for each $z \neq 1$ in G there is a homomorphism $\psi_z : G \to GL(m, E^{(q)})$ with $z^{\psi_z} \neq 1$, where m is an integer depending only on G and $E^{(q)}$ is the field defined in Lemma 10.4A. Since G is finitely generated, so is G^{ψ_z}. Since $E^{(q)}$ is algebraic, the entries of the matrices in G^{ψ_z} generate a finite field and so G^{ψ_z} is finite. The hypotheses now show that H^{ψ_z} is a finite nilpotent subgroup of $GL(m, E^{(q)})$. We assert that the class of nilpotency of H^{ψ_z} is bounded above by a constant independent of z. Since H^{ψ_z} is the direct product of its Sylow subgroups it is enough to show that the nilpotence classes of the finite p-subgroups of $GL(m, E^{(q)})$ are uniformly bounded for all primes p. For $p = q$, the class of a q-subgroup is $\leqslant m-1$ by Theorem 1.4A. For $p \neq q$ and $p > m$, the class of a finite p-subgroup is $\leqslant 1$ by Corollary 5.2. Finally, for $p \neq q$ and $p \leqslant m$, the order and hence the class of a finite p-subgroup is bounded by Lemma 10.4A (2). Thus we have shown that for some integer $c \geqslant 1$, the class of each finite nilpotent subgroup of $GL(m, E^{(q)})$ is $\leqslant c$. In particular, for each $z \neq 1$ in G, $\gamma_{c+1}(H^{\psi_z}) = 1$. Since H is isomorphic to a subgroup of the unrestricted direct product $\underset{z}{\otimes} H^{\psi_z}$ (see the note before Corollary 10.2), we conclude $\gamma_{c+1}(H) = 1$. Hence H is nilpotent. This completes the proof in the case where char $F \neq 0$.

Case 2 (char $F = 0$). Theorem 10.2 shows that for each $z \neq 1$ in G there is a homomorphism $\psi_z : G \to GL(m, E^{(q_z)})$ with $z^{\psi_z} \neq 1$, where m is an integer depending only on G and the prime q_z may be chosen arbitrarily provided it is larger than some bound. Using Dirichlet's theorem on primes in an arithmetic progression we may choose q_z to be of the form

$$q_z = 1 + (p_1 p_2 \dots p_s)^m + k(p_1 p_2 \dots p_s)^{m+1}$$

for some integer $k \geqslant 1$ where $p_1, p_2, \dots p_s$ are the primes $\leqslant m$. If we define a field $E^{(q_z)}$ as in Lemma 10.4A then the Lemmas 10.4A and 10.4B show that there is a uniform bound (independent of q_z) on the orders of the finite p-subgroups of $GL(m, E^{(q_z)})$ for all primes. Thus, in a manner similar to the proof in Case 1, we can conclude that there exists a constant c such that $\gamma_{c+1}(H^{\psi_z}) = 1$ for all $z \neq 1$ in G. Thus, as before, $\gamma_{c+1}(H) = 1$, and H is nilpotent as required. ■

We obtain an immediate corollary.

COROLLARY 10.4. *Let G be a finitely generated linear group. Then any locally nilpotent subgroups of G are actually nilpotent. In particular, all p-subgroups of G are nilpotent.* ■

§10.5 We now give some applications of Theorem 10.4.

THEOREM 10.5A. *Let G be a finitely generated linear group. Then the Frattini subgroup Φ(G) is nilpotent.*

Proof. We even prove the stronger result: the intersection H of all maximal subgroups of *finite* index in G is nilpotent. Since the Frattini subgroup of a finite group is nilpotent (see Huppert's book p. 270), it follows that for all normal subgroups N of finite index in G, $\Phi(G/N)$ is nilpotent. However, every maximal subgroup M/N of G/N contains HN/N and so $HN/N \subseteq \Phi(G/N)$. Thus HN/N is nilpotent for each normal subgroup N of finite index in G; so H is nilpotent by Theorem 10.4. ■

Note. The conclusion of Theorem 10.5A need not be valid if we drop the condition that G is finitely generated (see Exercise 1). Moreover examples of P. Hall show that, if we drop the condition that G is a linear group, then the Frattini subgroup of a finitely generated solvable group need not be nilpotent.

DEFINITION. We define the iterated commutators of elements x and y in a group G by

$$[x, y]_0 = x \quad \text{and} \quad [x, y]_k = [[x, y]_{k-1}, y] \quad \text{for } k = 1, 2, \ldots .$$

An element $y \in G$ is a *right Engel element* if for each $x \in G$ there is an integer k (depending on x) such that $[x, y]_k = 1$. In 1957 Baer proved (see Huppert's book p. 298):

In a finite group G, an element y is a right Engel element $\Longleftrightarrow y \in \mathrm{Fit}(G)$. In particular, if G is generated by a set of right Engel elements then G is nilpotent.

THEOREM 10.5B. *Let G be a finitely generated linear group. Then $y \in G$ is a right Engel element* $\Longleftrightarrow y \in \mathrm{Fit}(G)$. *In particular, if G is generated by a set of right Engel elements then G is nilpotent.*

Proof. If $y \in \mathrm{Fit}(G)$, then y is in normal nilpotent subgroup K of G by the definition of $\mathrm{Fit}(G)$. Then, for all $x \in G$, $[x, y] \in K$ and so $[x, y]_{k+1} = 1$ if k is the nilpotency class of K. Hence y is a right Engel element.

Conversely, suppose y is a right Engel element of G, and let H be the smallest normal subgroup of G containing y. We shall now show H is nilpotent and so $y \in H \subseteq \mathrm{Fit}(G)$. Indeed, for any normal subgroup N of finite index in G, Ny is a right Engel element in G/N, and HN/N is the smallest normal subgroup of G/N containing Ny. Therefore, by the theorem of Baer quoted above, HN/N is nilpotent. Then Theorem 10.4 shows that H is nilpotent as required. ■

COROLLARY 10.5B. *Let G be a linear group (which need not be finitely generated). If all elements of G are right Engel elements then G is locally nilpotent.*

In 1953 Gaschütz proved the following criterion for a normal subgroup of a finite group to be nilpotent (see Huppert's book p. 270):

A normal subgroup H of a finite group G is nilpotent if (and only if) $H/H \cap \Phi(G)$ is nilpotent.

THEOREM 10.5C. *Let G be a finitely generated linear group. A normal subgroup H of G is nilpotent if $H/H \cap \Phi(G)$ is nilpotent.*

Proof. Let N be a normal subgroup of G. Then each maximal subgroup of G/N contains $\Phi(G)N/N$ and so $\Phi(G/N) \supseteq \Phi(G)N/N$. Hence, if $x, y \in H$, then $x(H \cap \Phi(G)) = y(H \cap \Phi(G))$ implies $x(HN/N \cap \Phi(G/N)) = y(HN/N \cap \Phi(G/N))$. Thus there is a homomorphism from $H/H \cap \Phi(G)$ onto $(HN/N)/(HN/N \cap \Phi(G/N))$ given by $x(H \cap \Phi(G)) \mapsto x(HN/N \cap \Phi(G/N))$; since the former group is nilpotent, so is the latter. Thus the theorem of Gaschütz given above shows that, for each normal subgroup N of finite index in G, HN/N is nilpotent. Therefore H is nilpotent by Theorem 10.4. ∎

Exercises. 1. Let G be the subgroup of $GL(4, \mathbf{C})$ generated by

$$\begin{bmatrix} \cdot & 1 & & \\ & \cdot & 1 & \\ & & \cdot & 1 \\ 1 & & & \cdot \end{bmatrix} \quad \text{and} \quad \operatorname{diag}(\zeta, \zeta^2, \zeta^4, \zeta^3)$$

for all $\zeta \in \mathbf{C}$ such that $\zeta^{5^m} = 1$ for some integer $m \geqslant 0$. Show that $\Phi(G)$ contains all the diagonal matrices in G and is of index 2 in G. Prove it is not locally nilpotent.

2. An element x of a group G is a *left Engel element* in G if for all $y \in G$, $[x, y]_k = 1$ for some integer k depending on y. It is known that an element x is a left Engel element in a finite group $G \Longleftrightarrow x \in \bigcup_{k=1}^{\infty} Z_k(G)$ (see Huppert's book p. 298). Prove that the same condition holds when G is a finitely generated linear group.

§10.6 *Notes and References.*

The idea of 'finite approximation' was first introduced by Mal'cev in the paper quoted in §3.9. In particular, Corollary 10.2 is included there. The form in which the idea is developed in this chapter is essentially that of

Platonov, V. P. The Frattini subgroup of linear groups and finite approximability, *Soviet Math.* **7**, 1557-1560 (1966).

Theorem 10.5A is proved there.

The general results Theorems 10.3 and 10.4 have not previously been published. It now appears that several theorems on solvability or nilpotence of certain linear groups, proved by other means by various authors, are fairly direct corollaries of these results. Undoubtedly there are further general theorems of this type to be discovered.

The question of how some results can be extended to groups of matrices over Noetherian rings is considered in

Gruenberg, K. W. The Engel structure of linear groups, *J. Algebra*, **3**, 291–303 (1966).

Several recent papers of Wehrfritz deal with results like those of §10.5 by different methods. Other papers of interest include

Gruenberg, K. W. The hypercenter of linear groups, *J. Algebra*, **8**, 34–40 (1968).

Merzljakov, Ju. I. Verbal and marginal subgroups of linear groups, *Soviet Math.* **8**, 1538–1541 (1967).

Addendum

The references in the main body of these notes were based on my state of knowledge when the notes were completed at the end of 1968. In the meantime I have learned of further work which seems especially pertinent to the results described here.

A precise bound for $\beta(n)$ in Theorem 6.4 has been given in

Dornhoff, L. Jordan's theorem for solvable groups, *Proc. Am. math. Soc.* **24**, 533-537 (1970).

Dr. Michael F. Newman has informed me that there is an error in my paper on the solvable length of a linear group (see §2.11). But even so the final bound on the solvable length in the two main theorems remain valid. He plans to publish his corrected version soon, including an analysis of the Fitting length.

With respect to polycyclic groups (see §6.7) I should add a reference to the following notes which have recently been reissued and are available from the Canadian Mathematical Society, McGill University, Montreal:

Hall, P. The Edmonton Notes on Nilpotent Groups 1957.

The recent paper

Zassenhaus, H. On linear noetherian groups, *J. Number Theory*, **1**, 70-89 (1969).

has as its main theorem a conjecture of R. Baer that every noetherian linear group has a polycyclic subgroup of finite index. Unfortunately the proof contains a gap (see *Math. Reviews*, **39**, No. 4294 (1970)). It would be interesting to know whether the proof can be completed, or if indeed the theorem is true.

The reference to a paper of Dade in §5.8 should be supplemented by a reference to the review (*Zentralbl. Math.* **157**, 63 (1969)).

A substantial improvement to Theorem 7.3 has been announced by D. L. Winter (*Amer. Math. Soc. Notices*, **17** 648 (1970)).

Finally we mention that B. Wehrfritz has recently issued a set of interesting notes on Linear Groups available in the series of notes from Queen Mary College, London.

[*September* 1970]

INDEX